The Fiber Optic Association

Fiber To The Home Handbook

*For Planners, Managers, Designers, Installers And Operators
Of FTTH - Fiber To The Home - Networks*

Jim Hayes

The Fiber Optic Association, Inc.
The International Professional Association Of Fiber Optics
www.foa.org

The Fiber Optic Association

Fiber To The Home (FTTH) Handbook

The Fiber Optic Association, Inc.
Telephone: 1-760-451-3655 Fax 1-781-207-2421
info@foa.org   www.foa.org

Copyright 2021, The Fiber Optic Association, Inc.

ISBN: 9798541140118

The FOA logo, Fiber U® and CFOT® are registered trademarks
of The Fiber Optic Association, Inc.

Imprint: Independently published

Table of Contents

Chapter	Page
Preface	5
Chapter 1. Introduction to FTTH	7
Chapter 2 FTTH Architectures	19
Chapter 3. FTTH PONs: Passive Optical Networks	35
Chapter 4 FTTH Network Design	43
Chapter 5. Fiber To The Home Installation	67
Chapter 6. Testing FTTH Networks	77
Chapter 7. Case Studies - Do It Yourself FTTH	87
Chapter 8. FTTH Project Management	97
References	111

Preface

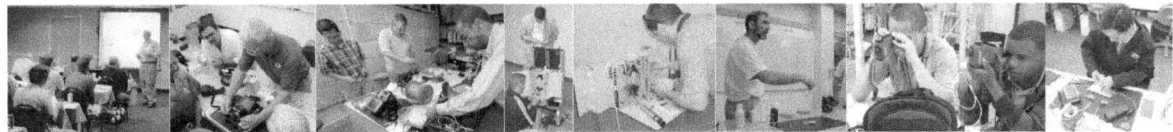

The purpose of this book is to provide a reference guide for anyone involved in a fiber to the home (FTTH) project, whether they are considering such a project in their area or planning, managing, designing, installing, operating, or using a FTTH network. The material in this book comes from FOA's involvement in many FTTH projects over the last two decades, consulting, creating technical material for reference and creating training programs for our online learning site, Fiber U (fiberu.org) and our approved schools around the world.

The Fiber Optic Association, Inc., the international nonprofit professional association of fiber optics, has become one of the principal sources of technical information, training curriculum and certifications for the fiber optic industry. As technology has driven the rate of technical change ever faster, it has become a challenge to provide reference books that are not hopelessly out of date.

The FOA created its online FOA Guide (www.thefoa.org/guide) to provide a more up-to-date and unbiased reference for those seeking information on fiber optic technology, components, applications and installation. Its success confirms the assumption that most users prefer the Internet for technical information.

FTTH - fiber to the home – has been a major application covered in the FOA Guide and the FOA CFOS/H one of the most popular FOA certifications. FOA Guide materials come from two decades of experience with FTTH including developing training curriculum for training techs for the earliest commercial installations of FTTH and consulting with many diverse FTTH projects.

This FOA FTTH Handbook is a compilation of all the FTTH materials from the FOA Guide with additional materials covering fiber optic project design and management. It may also be used as a reference for FOA CFOS/H FTTH Specialist Certification.

If you have feedback on the book, feel free to email comments or questions to the FOA at info@foa.org.

What is The FOA?

The Fiber Optic Association, Inc. is a international non-profit educational organization that is dedicated to promoting professionalism in the field of fiber optics. Founded in 1995 by a dozen prominent fiber optics trainers and industry personnel, it has grown to being the internationally recognized certifying body for fiber optic technicians.

Today FOA is involved in:
- Administering technical certification programs
- Evaluating and approving training schools
- Developing online & print technical references
- Developing curriculum for training
- Training instructors
- Participating in industry standards activities
- Publishing online and email newsletters
- Promoting fiber optics applications and education

The FOA has approved hundreds of training programs around the world, including those at technical high schools and colleges, union apprenticeships, professional societies, military groups, professional training organizations, service providers and fiber optic manufacturers and installers. FOA certification is recognized as the standard for fiber optic technician training and qualification by hundreds of organizations worldwide.

Certifications
FOA is an internationally recognized certifying body for fiber optic technicians. Fiber optic techs have achieved over 100,000 FOA certifications. FOA CFOT® certification is the most accepted certification for fiber optics techs worldwide. FOA CFOS/H certification for FTTH is widely accepted as the certification for FTTH designers, installers and operators.

FOA Course Approval and Curriculum Materials
The FOA approves training programs that meet its standards and those programs can offer FOA certifications to their students. For instructors teaching a fiber optics course, the FOA offers a complete curriculum package including Instructor's Guide, Student Guide, PowerPoint slides, instructions on setting up and running hands-on laboratories. Contact the FOA or go to the FOA website for more information.

A Note of Appreciation
This handbook has been produced and reviewed by a number of contributors whom we wish to thank for their work in contributing, creating and reviewing the materials included here including Tom Collins, Bill Graham, Gilberto Guitarte, Jerry Morla, Joe Botha, Ian Gordon Fudge plus Greg Turton, Southern FiberWorx, Kevin Woods, ConnectANZA, and others we may have overlooked.

This information is provided by The Fiber Optic Association, Inc. as a benefit to those interested in planning, designing, installing and operating FTTH fiber optic communications networks. It is intended to be used as a overview and/or basic guidelines and in no way should be considered to be complete or comprehensive. These guidelines are strictly the opinion of the FOA and the reader is expected to use them as a basis for learning, reference and creating their own documentation, project specifications, etc. The FOA assumes no liability for the use of any of this material. Those working with fiber optics in the classroom, laboratory or field should follow all safety rules carefully.

Chapter 1. Introduction to FTTH

Objectives: From this chapter you should learn:
What is fiber to the home or FTTH
What is FTTx or FTTH, FTTC, FTTP, FTTB, etc.
The importance of FTTH
Organizations that build and operate FTTH networks
The jargon of FTTH

Introduction To FTTH

The term FTTx is used as a catch-all for fiber to the home, premises, business and even the hybrid fiber to the curb. Mostly we're concerned with fiber to the home, but we will discuss all options. They all refer to technologies used to bring broadband to the user. By broadband we mean high speed Internet access and today there seems to be universal agreement that it is a necessary utility, not a luxury, for online education, entertainment and everyday life.

Broadband connections to the home began in 1997 with the installation of the first cable modems by CATV companies using their coax cable networks. Telcos were left to attempt to get high speeds over copper networks, but twisted pair wires were inferior to coax used by CATV. CATV became the dominant provider of broadband and the cable modem was adopted by the industry under the DOCSIS standard. In the years since, the backbones for both telco and CATV were upgraded to fiber, but even after the development of more than 20 generations of equipment, DSL (digital subscriber line) over twisted pair phone lines was clearly an inferior product. DSL continued until about 2020 when the last telco holdouts gave up.

Beginning around 2006, fiber to the home (FTTH) began emerging as a real application of fiber optics. Two factors made the cost low enough to justify replacing aging copper wires, the decline in fiber optic component costs following the fiber recession caused by the bursting of the Internet "bubble" in 2001 and the development of passive optical networks (PONs).

Fiber has now gained acceptance in the final frontier of telecom networks, the "last mile," the connection to the home. Many homes, apartments and businesses are still connected with aging, low performance copper telephone wire that cannot support connection speeds for broadband access. The costs of maintaining these old copper cable plants is also extremely high and always increasing. Even when telco landlines are abandoned for mobile phones, the home needs a connection capable of providing high speed Internet access and fiber is the most logical - and economic - choice, providing gigabit+ speeds with plenty of room to upgrade.

Phone companies, cities, utilities, commercial service providers and even real estate developers are now realizing the best choice for upgrading the subscriber connection is fiber to the home or premises (FTTH, FTTP) although fiber to the curb (FTTC) or fiber to wireless (FTTW) may still be used where appropriate. Wireless connections to the home using WiFi are being used in many rural areas where line-of-sight wireless links are feasible.

The possibility of delivering new services (the triple-play of phone, Internet and streaming video) and low-priced components for with new network architectures make FTTx financially attractive. Companies are spending billions of dollars connecting millions of homes and offices with fiber.

CATV companies who have used hybrid fiber-coax (HFC) networks in their backbones for decades even have their own standard for fiber to replace coax, since the cost of fiber is reasonable and performance unlimited. Municipalities or private individuals are installing their own FTTH systems when phone or CATV companies won't do it soon enough. Electrical and telephone coops are installing fiber in areas where the usual service providers won't because they don't see a large return on their investment. Housing developers are learning about FTTH because their customers are demanding the highest bandwidth broadband connections.

FOA's Role In FTTH
As the worldwide professional organization for fiber optics and the widely accepted certifying body for fiber optic techs, FOA has had a major role in FTTH development. We have been involved in developing training and certifying fiber optic techs for FTTH from the beginning.

In the process of developing standards for FTTH tech certification and creating reference materials for the FOA Guide and curriculum materials for our FOA-approved schools, we have developed an in-depth understanding of the technology and applications of FTTH. FOA has helped hundreds of organizations to design, install and operate FTTH networks - and in many cases help them make the decision to move ahead with the project by guiding them through technical, financial and logistical decisions that they needed to make.

All these plans depend on finding or training adequate numbers of technicians. Since being approached by Verizon to develop training for FiOS techs in 2006, FOA has been working with operating companies, municipalities, installers, and our approved schools to develop requirements for FTTx training and certification, with the goal of providing enough qualified FTTx installation technicians to make these plans possible.

FOA CFOS/H Certified FTTH Technician certification programs are now being taught in many FOA-approved schools worldwide. Students seeking certification must complete the FOA CFOT program first to obtain their first level certification, then attend a course on FTTx that will prepare them for the CFOS/H exam. Some techs attend courses for the FOA CFOS/D fiber optic network design certification also, since they are tasked with designing the networks.

Those interested in learning about FTTx but are not seeking certification will find the courses good information on the current technology. They can also take self-study courses on Fiber U, FOA's free online learning website.

A Short History Of Fiber Optics

Since the first installations of fiber optic networks in the late 1970s, the goal of the fiber optic industry has been to install fiber optics all the way to the home. Telecommunications systems were usually divided into long distance, metropolitan and subscriber categories. In the beginning, the challenge was well known - 10% of the cable plant was in long distance, 10% was in metro and fully 80% of the cable plant was the "last mile" - the subscriber. The issue was, and is, the economics of fiber to the home (FTTH.)

From an economic standpoint, fiber was immediately cost effective in the long-distance networks. Compared to copper or digital radio, fiber's high bandwidth and low attenuation easily offset its higher cost. Compared to copper wire used in telephony, fiber could carry thousands of times more phone conversations hundreds of times further, making the cost of a phone connection over fiber only a few percent as much as transmitting over copper. This photo was used many times in the early days to illustrate the information carrying advantage of optical fiber:

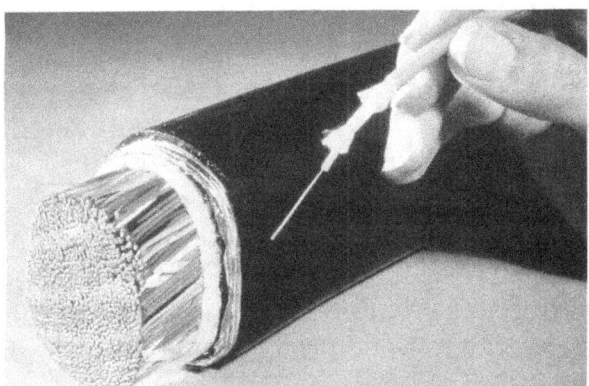

Photo from the 1970s by AT&T to show the relative information carrying capacity of copper and fiber cables of the era.

It took only a few years before the rapidly advancing technology of fiber optics led to widespread use and fiber quickly dominated the long distance market. Crews buried cables underground or ran aerial cables on poles nonstop for a decade to upgrade long distance service. At the same time, technology was developed for submarine cables and by the late 1980s, all overseas communications expansion was done by fiber optics, replacing copper cables and satellites. Today, virtually all long distance communications is carried over the installed fiber optic network both on land and via hundreds of submarine cables as shown below.

The next step was connecting local central offices, the link between subscribers and the switched phone network. Around the time the long distance networks were being completed, consumer use of the Internet took off. It was the Internet that drove the communications revolution by connecting anyone with a PC to a worldwide source of information and communication and forced the expansion of fiber into communications networks. Metropolitan phone networks became overloaded quickly and fiber optics was ready to provide the expansion capability. The scope of metropolitan fiber optic installations was obvious to anyone driving around almost any town, as it was hard to drive anywhere without encountering roads torn up for the installation of conduit and fiber optic splicing trucks blocking the roadways.

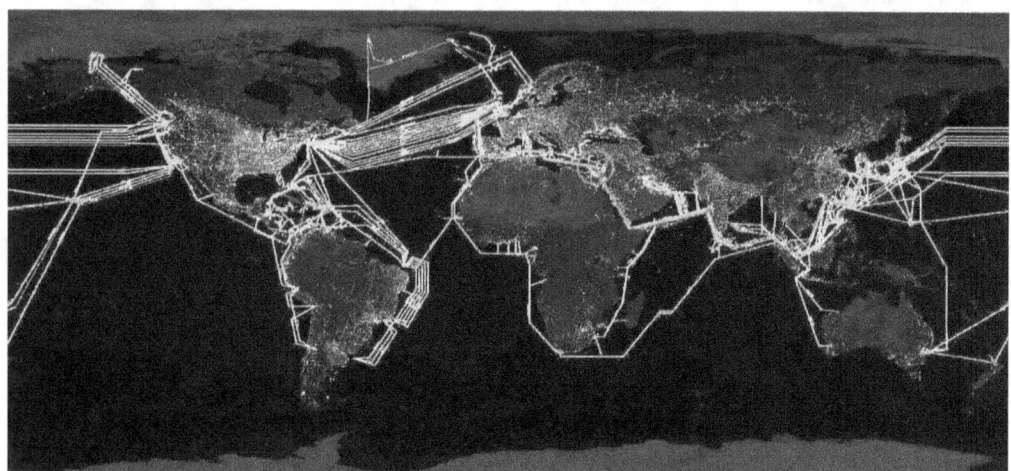
The continents are connected with undersea fiber optic cables

While telecom companies were building out their backbone networks, CATV companies benefited from several technical developments. The Internet created a need for high speed "broadband" networks. A way of using CATV networks for Internet access using the frequencies from several TV channels was developed leading to the cable modem. And finally, the DFB laser capable of converting CATV analog signals to optical was developed to allow companies to build a hybrid fiber coax (HFC) CATV network. CATV companies ran with the idea and began offering broadband Internet service in 1997, jumping ahead of the telecom networks to become the dominant broadband providers.

Then the telecom/Internet "bubble" burst in 2001. The Internet "bubble" that caused the telecom "bubble" and thereby the fiber optic "bubble" caused the downfall of a tremendous number of companies and left the industry with a glut of both installed fiber backbone capacity and fiber optic component manufacturing capacity. In a good illustration of economics at work, the cost of fiber optic components took a nosedive as supply outstripped demand. Since the market bust, fiber and component prices became very cheap. One analyst compared fiber prices to kite string and fishing line, both of which are more expensive than the current prices of top-quality singlemode optical fiber. That set the stage for the next big application for fiber optics.

Many homes today are still connected with aging, low-performance copper telephone wire that cannot support DSL connection speeds that allow the phone companies to compete with the cable modems used by CATV companies for broadband access. These aging phone lines not only cannot carry high bandwidth digital signals, they are extremely expensive to maintain just for POTS (plain old telephone service.) Savings in maintenance alone were projected by a 2005 Telcordia report (the Bernstein Report) to pay back the cost of installing fiber in under 20 years, irrespective of revenue from new services.

One problem with converting homes from copper to fiber is the immense size of the task. Long distance cables represent about 10% of the telecom network. Metropolitan networks represent another 10%. But the connection to the home, traditionally called the "last mile," represents about 80% of all the cabling in telecom, making conversion of copper to fiber to the home a massive task.

Besides component prices dropping because of oversupply, new network architectures have been developed that allow sharing components for FTTH that further reduces cost. A passive optical splitter that takes one fiber input and connects bi-directionally to as many as 32 users (more in some versions, up to 256 users) cuts the cost of the links substantially by sharing, one expensive laser transmitter in the central office with 32 or more homes and reducing the number of fibers needed. This is what is called a PON network, a passive optical network.

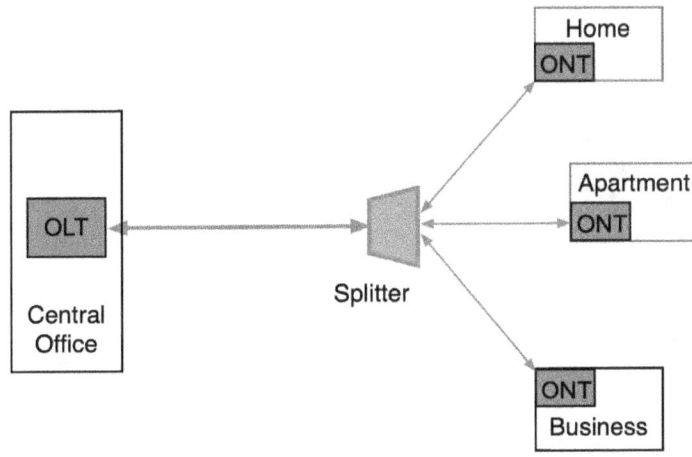

FTTH PON block diagram

Each home needs to be connected to the local central office or head end with singlemode fiber through a splitter generally placed close to the homes connected to it. Every home will have a singlemode fiber link placed underground or aerial to the cables running down the street and a network interface device containing fiber optic transmitters and receivers will be installed on the outside of the house or indoors. The incoming cable needs to be terminated at the house, tested, connected to the interface and the service tested.

Now fiber has gained acceptance in the final frontier of telephone networks, the "last mile"—the connection to the home. Phone companies are now realizing the only choice for upgrading the subscriber connection is fiber to the home (FTTH). Service providers have committed billions of dollars to connecting millions of homes with fiber.

With FiOS in 2006-7, Verizon became the leader in the US in FTTH, but the US trails many other countries around the world in converting to FTTH. Progress requires massive capital investment and training lots of people to install FTTH, or FTTP (fiber to the premises as they call it.) Besides the telcos, several other groups are installing FTTH. Even the CATV companies are considering fiber to replace aging coax to the home, since the price is right and performance unlimited. The most common exception to FTTH are rural areas where FTTH is too expensive, such as rural areas, wireless or satellite connections may be used instead.

National Telecommunications Companies: Countries with a national telecom policy usually favor FTTx for their national broadband networks. In 2021, the highest percentage of users connected on FTTH was in UAE where Etisalat has connected virtually all homes (Etisalat is an FOA approved school also.)

FTTH optical network terminal

Municipalities: Some of the first FTTH systems were installed by cities - progressive ones like Palo Alto, CA did it at the request of their high-tech citizens, some did it to entice businesses to move there, like Anaheim, CA some did it (or are trying to) because they were not pleased with the service of telcos or CATV companies. The latter often found the telcos or CATV companies to be formidable opponents who did not always play fair! Most municipal FTTx projects use rights of way available to the city through city-owned utilities such as Chattanooga, Clarksville and Jackson, TN which offer 1Gb/s FTTH already.

Google Fiber made municipal FTTH popular by having a competition for a city to get Gigabit FTTH installed by them - and Kansas City won the competition, but many other cities decided they wanted fiber optics anyway and they are now running their own networks. Google fiber had many successes and some notable failures, but they convinced many other organizations to commit to gigabit broadband over fiber.

Utilities: Owning rights of way to the home convinced some utilities to try FTTH or FTTC. Ethernet over power lines was once considered an option for power companies ton use power lines for the final connection to the home, but the technology was inadequate. FTTx is even becoming real for many customers through electrical cooperatives. The best example of a utility owned FTTH network is the EPB system in Chattanooga, TN.

Private companies: There are private companies that will build municipal FTTH networks under an agreement with the city, similar to CATV service provider agreements. In addition, some contractors and real estate developers building subdivisions or apartments are installing FTTH so they can connect with telecommunications companies for services to resell.

Google Fiber is certainly the largest and best known of the private FTTH networks. Google started in Kansas City, then added Austin and Provo, Utah and began adding up to 34 other cites to their service areas. What Google appears to have learned from Kansas City was when the city is cooperative, FTTH is relatively cheap to install (KC numbers have been quoted as low as <$500US per customer including installation) and provides a constant stream of high margin income.

As of 2021, there are thousands of FTTH projects around the world. Many are being done by telecom companies like Etisalat. Cities, counties and even states are building fiber optic networks to serve citizens. Real estate developers are building FTTH networks because the cost is lower than a upgraded kitchen, helps sell houses faster and brings much more profit. Even venture capital has begun investing in FTTH networks for the potential return on investment.

What Makes A Successful Fiber Optic Project?
People call FOA for advice all the time. Most of the calls deal with technical questions about products, installation, and testing. But in one call; a manager who was starting to plan a fiber optic project wanted advice on how to proceed. It was a long call! His basic question was "What does it take to have a successful fiber optic project?" We responded with 4 words: financing, commitment, expertise, and patience.

Financing: The story goes that someone asked Neil Armstrong what he was thinking about while sitting on top of the rocket ready to launch Apollo 11 to the moon. "Every part was made by the lowest bidder," was supposedly his reply. (The same quote has been attributed to most early astronauts!)

Fiber optics are not necessarily expensive; in fact, fiber has been used so widely because it is the least expensive communications medium in virtually all projects. But fiber optic projects may require a lot of construction which makes the project expensive. Like all other projects, it never pays to cut corners. Planning and running the project properly is what saves money, trying to cheapen the project. Not all jobs should go to the lowest bidder, unless they meet all the criteria for a qualified bidder. Likewise, one needs to ensure that when a project starts, there are funds available to complete the job properly, including some extra for unplanned changes or modifications.

Commitment: Just like having sufficient finances to compete the project, one needs a commitment to finish the job once it is started. Changes of management or changes in governments often lead to confusion or even modifying a project in midstream. There is nothing wrong with making changes based on what learns as the project progresses; it may even involve greater efficiency or cost savings, but arbitrary changes may jeopardize the project's timetable, completion or even its usefulness.

If the project is under the auspices of a government entity, changes in administration or management that causes changes in a project will invariably make it more expensive and may jeopardize the success of the entire project. Ideally, the personnel who propose, design and plan the network should see it to completion.

Expertise: Fiber requires expertise and experience. It's obvious the installers need to know what they are doing, but in reality, so must the managers who work for the organization that is contracting for the work. There are many instances of projects where the managers signed off on the project when it was incomplete or improperly installed. The only way to properly manage a project is to understand every aspect of it well enough to know if it is being done properly and when it is actually complete.

Planners, designers, contractors and installers should all be trained and certified as well as being experienced with good references. That holds doubly so for consultants. In many places, to be a consultant or cabling contractor means little other than registering as a business and advertising your services. Some of the problems we've seen with outside services, include consultants who took contracts, spent time on a project, then told the customer they could not help them with the project, but kept the money.

We have seen contractors doing shoddy installations, ruining expensive fiber optic cable during pulling and leaving jobs half done but getting paid because the customer knew no better. One of the biggest problems is subcontractors. A contractor with good credentials gets the job but subcontracts some of the work to a contractor who will do the work at a lower price, but does not have the training or experience (or motivation) to do it right. In your contract with an installer, we recommend a clause giving the project manager responsibility for evaluating and approving all subcontractors.

The manager must know better to prevent problems. The final chapter on this book covers network planning and management. FOA also has pages in the FOA Guide on what the manager needs to know.

Patience: From concept to acceptance, a typical OSP fiber project can take 2-5 years and a premises project 1-2 years. It depends on the size of the project, the time to properly design it, create project paperwork, get permits, buy components, hire contractors and properly install it. Proper workmanship takes time and is not easily rushed. Saving time generally means cutting corners and that is often the cause of the problems encountered. Take your time, plan, design, select, install, test and document your network properly.

And by the way, "future proofing" is a myth! Who would have known in 1990 how ubiquitous the Internet would be today? How reliant we could be on smartphones other mobile devices? How many workers would be working remotely or using videoconferencing for meetings? Technology moves too fast and is too disruptive for anyone to make reliable predictions. The IBMer who developed MRP - the original company organizational software - used to tell everyone, "A forecast is wrong from the moment it is made." Plan for the future, but assume you will upgrade, change directions, etc. driven by new tech and changes in the world around us.

FTTH Jargon

The key to understanding any technology is understanding the language of the technology – the jargon. Here is an overview of some FTTH and fiber optic jargon to introduce you to the language of fiber optics and help you understand what you will be reading in the book.

Read this section to help your understanding of the rest of the book and refer to it when you encounter a term that you do not recognize. If you are not familiar with fiber optics in general, you can also use the FOA Online Reference Guide for more in-depth explanations. (https://foa.org/guide/)

FTTH, Fiber to the Home: Connecting broadband Internet subscribers with fiber optics all the way to the home.

FTTx: The term FTTx is used as a catch-all for fiber to the home (FTTH), premises (FTTP), business (FTTB) and even the hybrid fiber to the curb (FTTC). Mostly we're concerned with fiber to the home, but we will discuss all options in the chapter on FTTH architecture.

Fiber, optical fiber, fiber optics: FTTH is "fiber" to the home and is based on fiber optics. Optical fiber used in FTTH is an extremely pure glass fiber the size of a human hair that transmits light through it's core which is $1/10^{th}$ the size of the fiber itself. FTTH uses conventional singlemode fiber standardized as G.652 fiber or its version with lower bend sensitivity, G.657.

PON, Passive optical Network: A network architecture used for FTTH and sometimes enterprise LANs that uses bi-directional optical splitters instead of switches to connect multiple users to one port of a central office or head end network terminal. PONs come in several varieties.

APON: The first PON standard, APON, was quickly replaced by BPON because it had no provision for broadcast video and digital TV was several years away.

BPON: The original PON standard based on older telco protocols (ATM) and analog (AM) video.

GPON: Gigabit PON is the newer, most popular PON version based on IP (Internet protocol). GPON has speeds of 2.5 Gb/s downstream to users and 1.25 Gb/s upstream from the user to the OLT. A 10 Gb/s version (10GPON) has been standardized with a unique update possibility, it can be installed on an operating GPON network and operate simultaneously with GPON, offering the possibility of having both a low cost gigabit consumer network and a 10G commercial network.

EPON: Ethernet PON is based on Ethernet for the first mile.

OLT: Optical Line Terminal, the equipment at the central office or head end that converts an Internet connection into PON protocols and transmits signals from a single fiber optic port to multiple users through optical splitters in the PON cable plant. The OLT connects to the Internet through a router to the service provider.

ONT: Optical network terminal, the equipment located at the subscriber that receives transmissions from the OLT and communicates back to it through the splitters of a PON fiber optic cable plant. Sometimes also called ONU, optical network unit.

Splitter: An optical component that takes one optical fiber input and splits it into multiple optical fibers on the output. Splitters come in numerous configurations, but in PONs the usual splitter has one or two inputs (a second for a spare or network monitor) and 2, 4, 8, 16 or 32, 64, 128, etc. outputs. Splitters work in both directions, so an OLT port can connect through a single fiber to a splitter with multiple outputs, each on a single fiber also, that connect to the user ONT. The signal from the OLT is split and sent to all users' ONTs (see encryption below) and signals from the ONTs are combined in the splitter into the single fiber connecting to the OLT port.

Encryption: Since the OLT transmission goes to all users at once, each user has a special encryption code to decode only their data, maintaining privacy. Encryption has made PONs popular with users that want to maintain privacy like government agencies.

WDM, wavelength division multiplexing: Since a PON sends signals in both directions on a single fiber, the downstream signals are sent at one wavelength (1490 nm) and upstream at another wavelength (1310 nm) to prevent signal interference. BPON used a

3rd wavelength, 1550 for downstream analog video. The new 10 gigabit PONs use slightly different wavelengths to allow simultaneous operation over the original PON like GPON.

MDU, multi-dwelling unit: The term used for building with multiple units to be connected with a FTTH network. This encompasses apartments or flats, condominiums or even multiple small businesses. MDUs have special requirements for FTTH networks running cables in the building and placing equipment like splitters and ONTs.

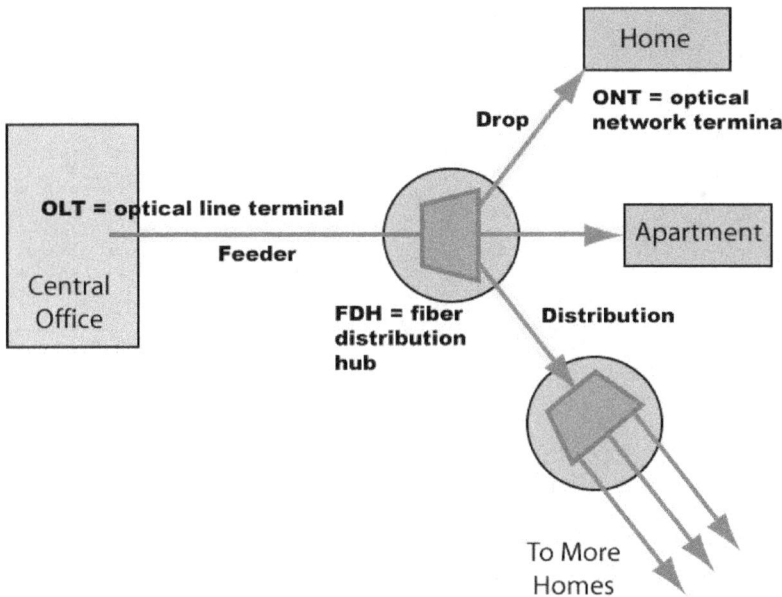

FTTH jargon for typical PON network components

Feeder cable: The fiber optic cable that goes from the central office or head end to the fiber distribution hub (FDH) housing the splitters.

Distribution cable: The cable from the fiber distribution hub (FDH) to the point that the single cable to the user, the drop cable, will be connected.

Drop cable: The small cable with 1 or 2 fibers that connects the user to a fiber in the distribution cable.

FDH, Fiber distribution hub: In networks with splitters located near the users, the FDH will house the splitters and connecting fibers for each user. The FDH may be a pedestal or a specialized splice closure with room for splitters.

Prefab, preterminated cables: Many FTTH networks are being built with drop cables already terminated with connectors. This allows the tech doing the installation at the home to work much faster and more efficiently than splicing connectors on to the drop cables. Some cables will only be terminated at the user end and the other end is spliced or terminated the drop point by an experienced fiber splicer.

Chapter 2 FTTH Network Architectures

Objectives: From this chapter you should learn:
The differences between fiber to the home architectures
Why fiber to the home is the ultimate solution
Why PONs (passive optical networks) are the most popular FTTH networks
How FTTH is used in single and multiple dwelling units

Fiber To The Home Architectures

New network architectures have been developed to reduce the cost of installing high bandwidth services to the home, often lumped into the acronym FTTx for "fiber to the x". These include FTTC for fiber to the curb, also called FTTN or fiber to the node, FTTH for fiber to the home and FTTP for fiber to the premises, using "premises" to include homes, apartments, condos, small businesses, etc. Recently, we've even added FTTW for fiber to wireless.

Let's begin by describing these network architectures.

FTTC: Fiber To The Curb (or Node, FTTN)

Fiber to the curb brings fiber to the curb, or just down the street, close enough for the copper wiring already connecting the home to carry DSL (digital subscriber line, or fast digital signals on copper.)

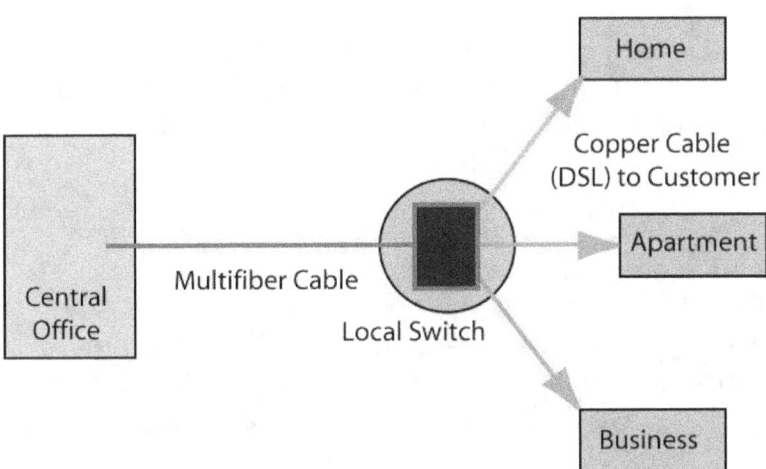

FTTC – fiber to the curb

FTTC bandwidth depends on DSL performance where the bandwidth declines over long lengths from the node to the home. There have been many types of DSL (ADSL, HDSL, RADSL, VDSL, UDSL, etc. - over 22 varieties in all) that offer varying performance over length, including some which "bond" more pairs of wires to improve the bandwidth.

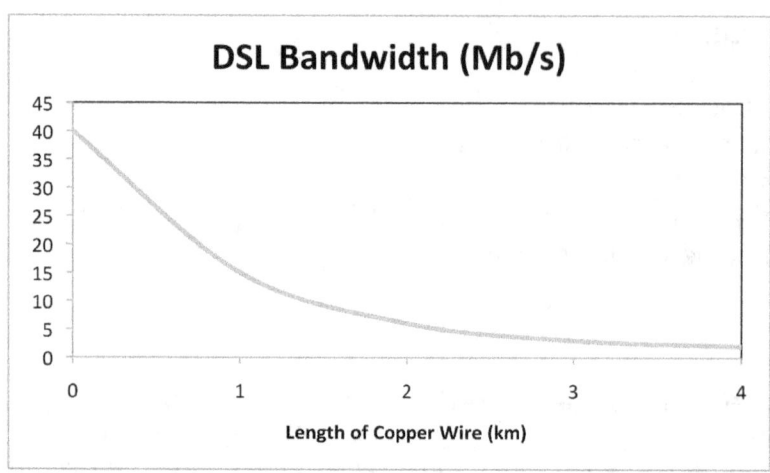

DSL bandwidth is a function of copper cable length

Newer homes that have good copper and are near the DSL switch can expect good service up to about 20 Mb/s. Homes with older copper or longer distances away will have less available bandwidth.

FTTC is less expensive than FTTH when first installed, but since performance depends on the quality of the copper wiring currently installed to the home and the length to reach from the node to the home, the level of service may be obsoleted quickly by customer demands. In older areas where the copper wires are of poorer quality or have degraded over time, DSL is difficult or impossible to implement and very expensive to maintain.

While there are still many DSL subscribers, by 2020 service providers basically abandoned it as obsolete. Now some large service providers who offer both landline and wireless are proposing using 5G wireless for the drop to the home. See below.

FTTW: Fiber to Wireless
Of course, today's mobile device users depend on wireless connections for their laptops, smartphones and tablets. Even many homes and businesses are now using wireless connectivity, especially those outside areas where FTTH or FTTC are not available or considered economical for future installations. Options for wireless include cellular systems which are the most widely available wireless solution around the world,

WiFi which has become available inside many businesses and even outdoors in areas served by municipal networks and satellite wireless, is used in many rural areas where distances are so large that cabling or WiFi is unfeasible. Options are primarily 5G since other proposed systems like WiMAX and Super WiFi, land-based wireless with longer ranges and higher bandwidth capability than most cellular systems have not been accepted. WiFi 6 will probably have higher bandwidth than 5G since it does not have the overhead to ensure cell to cell mobility.

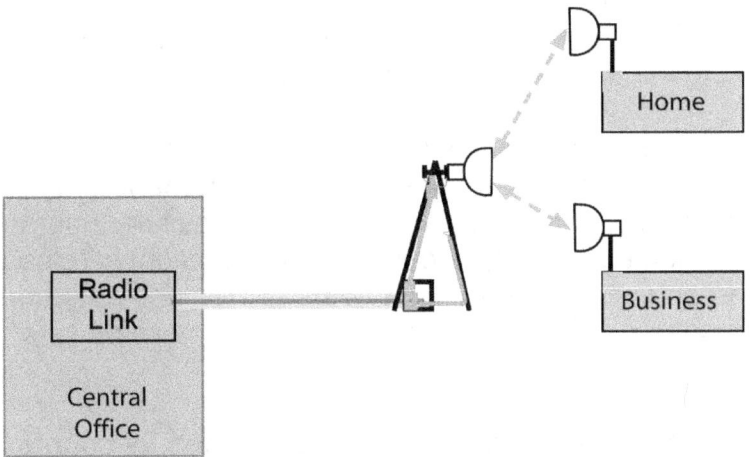

FTTx Wireless line-of-sight WiFi or other wireless networks can be used where cables are too expensive to install.

Small cells like this one are being installed in many cities around the world. They can offer better cellular service and perhaps with the introduction of 5G offer broadband type Internet speeds.

Small cell antennas with more localized coverage like the original small cell Light Cube Radio introduced by Alcatel-Lucent several years ago can be placed anywhere and connected with fiber and power. Small cells with 5G service are claimed to be capable of providing more bandwidth to users more efficiently. That assumes the 5G antenna is working at millimeter wave frequencies, not current cellular frequencies, but mm wave radio waves can have a problem penetrating walls, glass, trees and leaves, clouds, fog, snow, rain, etc.

Their claim is that 5G offers enough bandwidth to compete with fiber optics, but 5G has a problem getting inside buildings. A 5G home Internet connection places an antenna outside the building, comes into the building on wires and has a module inside the building to provide wired and wireless connectivity. As we write this, service providers have just started promoting 5G Internet, so its performance is unproven.

All these wireless systems depend on the same fiber optic communications backbones that everyone else does. As they grow, higher bandwidth demands means more traffic to local antennas which makes fiber more attractive. Most cellular users are converting older antenna towers connected by copper cables or line-of-sight wireless over to fiber. Fiber is even being used for connections up towers to wireless antennas as it is smaller and lighter than the coax cables previously used. Read more on how wireless depends on fiber here.

Wireless antennas require lots of fiber to carry the data to the antenna, of course, but also require power for the electronics and service calls for upgrades, something a PON does not require.

The biggest drawback to wireless Internet has been the cost of cellular service. Customers who want to download HDTV to watch at home will find generally wireless connections prohibitively expensive, but 5G may change that.

FTTH Active Star Network

The simplest way to connect homes with fiber is to have a fiber link connecting every home to the phone company switches, either in the nearest central office (CO) or to a local active switch.

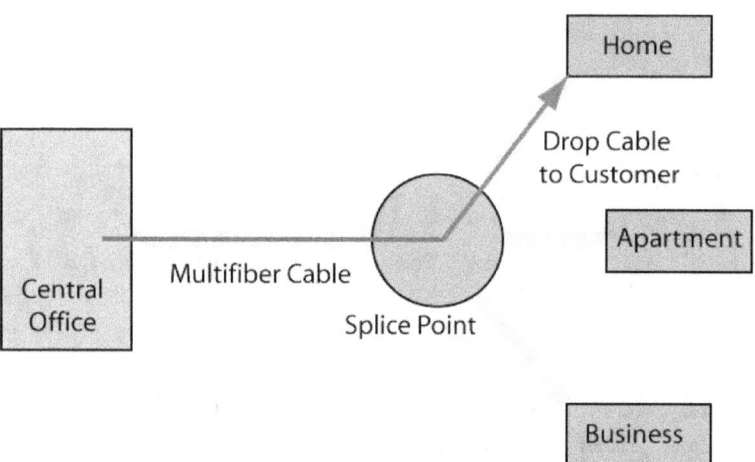

FTTH home run connects every subscriber directly to the central office
The drawing shows a home run connection from the home directly to the CO, while below, the home is connected to a local switch, like FTTC upgraded to fiber to the home.

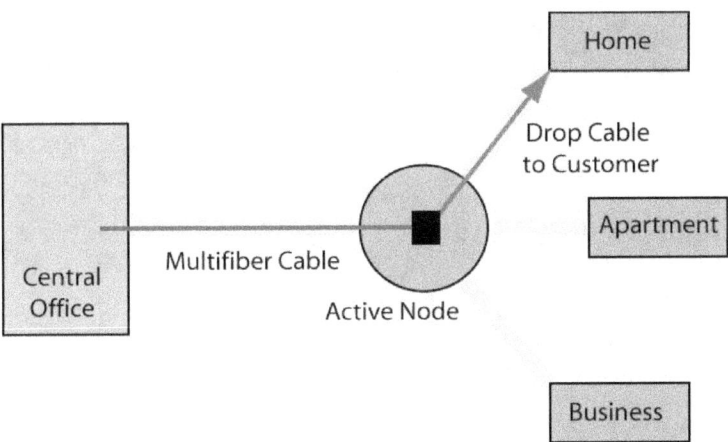

Active star architecture is similar to FTTC but with fiber connecting the home

A home run active star network has one fiber dedicated to each home (or premises in the case of businesses, apartments or condos.) This architecture offers the maximum amount of bandwidth and flexibility, but at a higher cost, both in electronics on each end (compared to a PON architecture, described below) and the dedicated fiber(s) required for each home.

FTTH PON: Passive Optical Network

PONs or passive optical networks have become the most popular design for FTTH networks because of their advantages in initial and operating costs. A PON system allows sharing expensive components for FTTH. A passive splitter that takes one input and splits it to broadcast to many users cuts the cost of the links substantially by sharing, for example, one expensive laser with 32 or more homes. PON splitters are bi-directional, that is signals can be sent downstream from the central office, broadcast to all users, and signals from the users can be sent upstream and combined into one fiber to communicate with the central office.

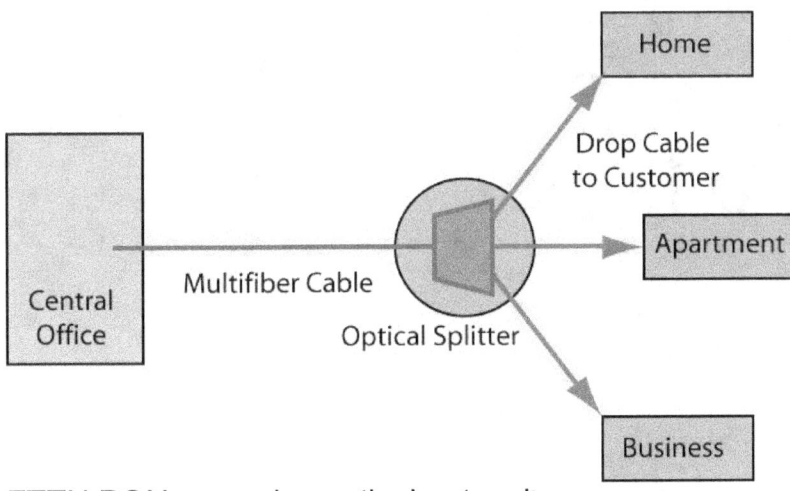

FTTH PON – passive optical network

Because of all the splitters and short fiber optic links, plus since some systems are designed for AM video like CATV systems, non-reflective connectors (like the SC-APC angle-polished connector) are generally used.

PON Splitter Ratios And Losses

Splitter Ratio	1:2	1:4	1:8	1:16	1:32	1:64	1:128
Ideal Loss / Port (dB)	3	6	9	12	15	18	21
Excess Loss (dB, max)	1	1	2	3	3	3	3
Loss (dB)	4	7	11	15	18	21	24

The splitter can be one unit in a single location as shown above or several splitters cascaded as shown below. Cascaded splitters can be used to reduce the amount of fiber needed in a network by placing splitters nearer the user. The split ratio is the split of each coupler multiplied together, so a 4-way splitter followed by a 8-way splitter would be a 32-way split. Cascading is usually done when houses being served are clustered in smaller groups. Splitters are sometimes housed in the central office and individual fibers run from the office to each subscriber. This can enhance serviceability of the network since all the network hardware is in one location at only a small penalty in overall cost for either dense urban areas or long rural systems.

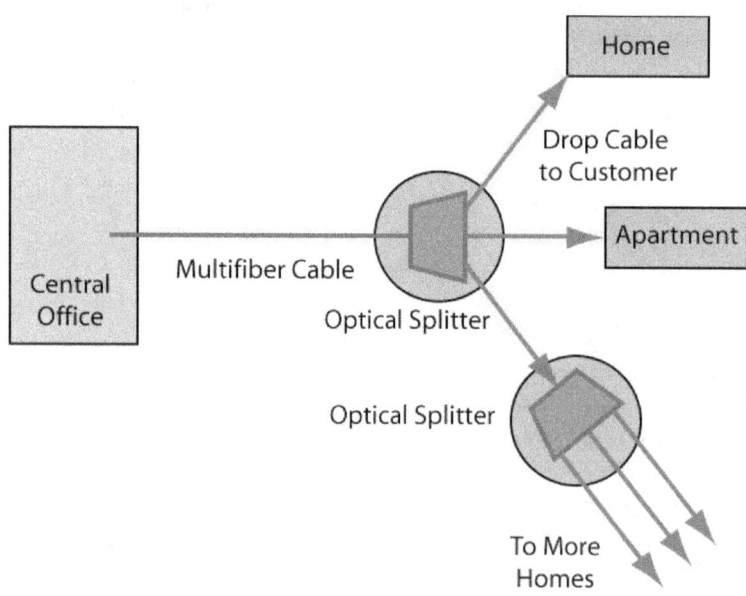

FTTH PON cascade

Most PON splitters are 1X32 or 2X32 or some smaller number of splits in a binary sequence (2, 4, 8, 16, 32, 64, 128, etc.). Couplers are generally symmetrical, say 32X32, but PON architecture doesn't need but one fiber connection on the central office side, or sometimes two so one is available for monitoring or as a spare, testing and as a spare, so the other fibers are cut off. Couplers work by splitting the signal equally into all the fibers on the other side of the coupler, Splitters add considerable loss to a FTTH link, limiting the distance of a FTTH link compared to typical point-to-point telco link. When designing a fiber optic network, here are guidelines on loss in PON couplers.

Each home needs to be connected to the local central office with singlemode fiber through an optical splitter. Every home will have a singlemode fiber link pulled into underground conduit or strung aerially to the phone company cables running down the street. Verizon has pioneered installing prefabricated fiber links that require little field splicing.

PON preterminated aerial installation

Here is a fiber distribution system that has been spliced into cables connected to the local central office. The preterminated drop cable to the home merely connects to the closure on the pole (arrow) and is usually lashed to the aerial telephone wire already connected to the home.

If the cable is underground, it will usually be pulled through conduit from connection to the distribution cable or the splitter to the home. Here a preterminated systems has two home drops connected to the distribution cable.

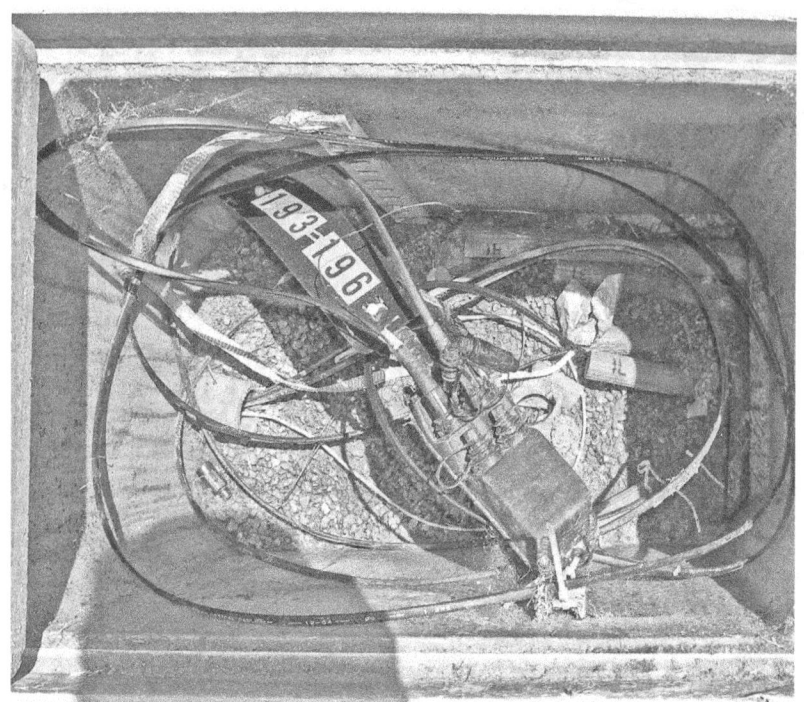
PON underground

The splitter can be housed in a central office with a fiber run to each subscriber (centralized splitter) or in a FDH – fiber distribution hub – housed in a pedestal in the neighborhood near the homes served or in a splice closure used to connect drop cables. The advantage of PONs is that this FDH is passive - it does not require any power as would a switch or node for fiber to the curb.

Fiber distribution hub (FDH) in a pedestal in Dubai

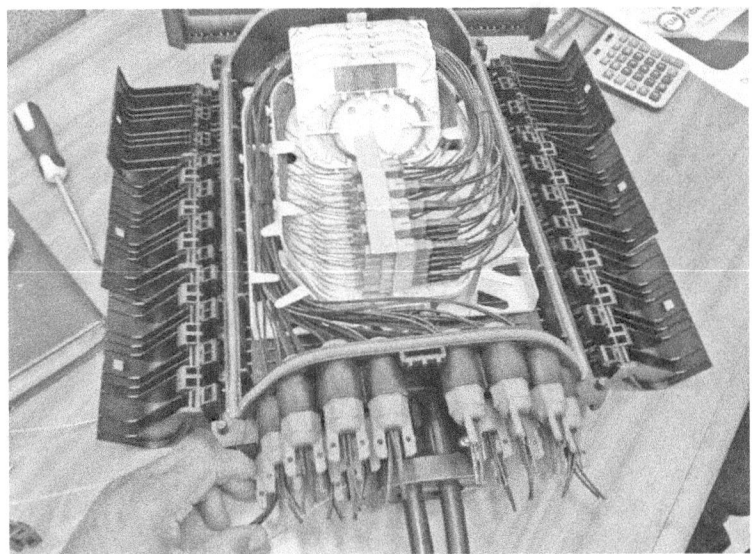
Splice closure with splitters and connections for prefab drop cables

A network interface device containing fiber optic transmitters and receivers may be installed in a box on the outside of the house or inside where the ONT looks like a cable modem or DSL modem. The incoming cable needs to be terminated at the house, tested, connected to the interface and the service tested.

FTTH ONT outdoors on wall of house

Below is the layout of a typical PON network with the equipment required at the CO, fiber distribution hub and the home. This drawing shows the location of the hardware used in creating a complete PON network. This drawing also defines the network jargon for cables: a "feeder" cable extends from the OLT (optical line terminal) in the CO (central office) to a FDH (fiber distribution hub) where the PON (passive optical network) splitter is housed. It then connects to "distribution" cables that go out toward the subscriber location where "drop" cables will be used to connect the final link to the ONT (optical network terminal).

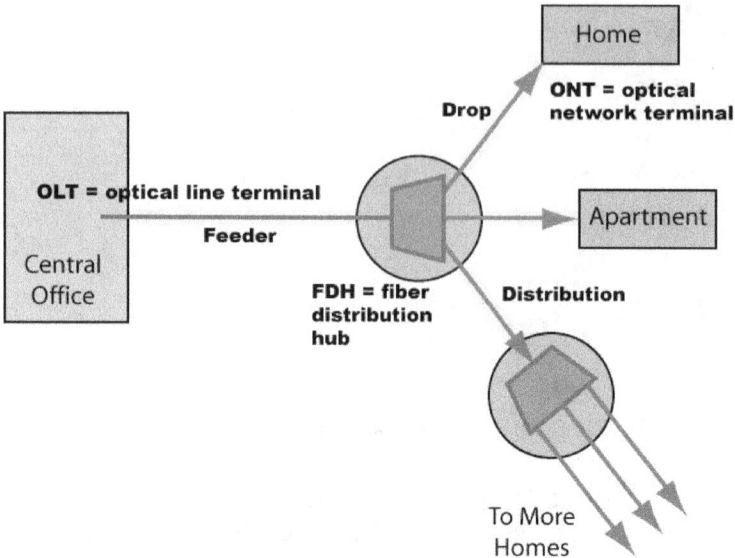

FTTH jargon for typical PON network components

Triple Play Systems

Most service providers' FTTH systems are "triple play" systems offering voice (telephone), video (TV) and data (Internet access.) To provide all three services over one fiber, signals are sent bidirectionally over a single fiber using two or three separate wavelengths of light. Three different protocols have been standardized, BPON, shown below, was the first system used but now mostly obsolete, used a third wavelength for AM video, while EPON and GPON use digital IPTV transmission. Read more on PON protocols.

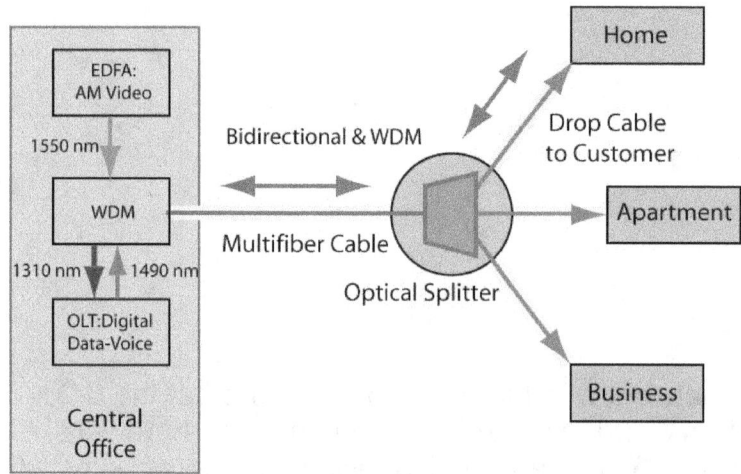

FTTH BPON wavelengths with analog TV at 1550nm

Downstream digital signals from the CO through the splitter to the home are sent at 1490 or 1550 nm. This signal carries both voice and data to the home. Video on BPON systems used the same technology as CATV, an analog modulated signal, broadcast separately using a 1550 nm laser which might require a fiber amplifier to provide enough signal power to overcome the loss of the optical splitter. GPON and EPON use digital IP TV for video. Upstream digital signals for voice and data are sent back to the CO from the home using an inexpensive 1310 nm laser. WDM couplers separate the signals at both the home and the CO.

Connecting To The Internet

Since the major goal of FTTH or any ot the other FTTx architectures is to provide high speed access to the Internet, the system operator must themselves have a high speed connection to the Internet backbone. In a PON architecture, it is the OLT that connects to the Internet through a router to the service provider. While planning a system, it is important to identify sources of Internet services, determine their location and how they would provide a connection to your system, and of course, the cost, which will depend on your location and how fast a connection you need.

Being a network, the PON or other network shares bandwidth among all its users. If 32 users connect to an OLT through a splitter, they will share the bandwidth available from the OLT port. Then the users attached to all the OLT ports share the Internet backbone connection incoming to the whole FTTH network.

A common question asked is how fast the Internet backbone connection should be. That's not an easy question to answer because it depends on how many users there are, what services are provided and the time of day. Video uses much more bandwidth and more people tend to watch video in the evening, but video conferencing will be more prevalent during the day. And bandwidth requirements are increasing because of new services that use more bandwidth.

Our recommendation is that you contact Internet service providers when planning a system and get their advice and estimates of costs.

Powering FTTH

Traditionally, telephone services, at least what are called "POTS" or plain old telephone service, have been self-powered from the central office. POTS phones were on a current loop powered from batteries or some other type of uninterruptible power in the CO. When a subscriber had an electrical power outage, they expected to be able to still use their phone, to call the electrical utility to report the outage, of course! Obviously, FTTH is not going to operate the same way. Fiber does not easily deliver electrical power, although systems have been developed to power sensors over light in the fiber, it is inefficient and expensive. Many FTTH systems provide a battery backup at the customer premises powered from the customer electrical system to keep the system operational during power outages. Some systems use the old copper wires replaced by the fiber to deliver power to keep the backup charged, so that the FTTH system provider pays for the power needed by the system. And some systems, recognizing that most people have a mobile phone, do not address the issue of backup power at all.

Urban/Suburban/Rural

Geography plays a big part in the design of a FTTH network, mainly in how it determines subscriber density. Dense population areas require less cable and generally higher fiber splitters, suburban areas with lower density often use cascaded splitters to serve few subscribers per splitters and rural areas often require long cable runs and decisions on whether to use fiber or wireless to connect the subscriber. Rural networks have several different options including taps for splitters and remote OLTs. We will discuss this in more detail in the FTTH Design page.

FTTH in MDUs (Multiple Dwelling Units)

When we normally talk about FTTH, we assume we are installing the fiber to a "home" where it terminates in a optical line terminal (OLT) and services (voice, data and video) are delivered inside the subscriber's home. But since we may have detached single-family homes, row houses or units in a large building, the situations can be quite different, requiring different architectures and installation practices. We should add that office buildings are often similar to MDUs, with the exception that floor plans are generally more flexible and units are larger, but the concepts are similar.

Let's assume we have fiber to the building, then what's next? We must decide how to deliver broadband to each unit in the building and then inside the unit. What are the options for delivering services to each unit from the entry facility?

Options for connecting each unit in the building include using:
- Currently installed phone lines using xDSL technology
- Currently installed CATV or satellite coax using cable modem or MOCA (Multimedia Over Coax Alliance) technology
- Installing wireless access points at appropriate points in the building connected by Ethernet as is usually done in hotels
- Installing new category rated UTP cable to each unit if it is within the 100m distance limit and use Ethernet
- Installing new coax cable to each unit and use cable modem or MOCA
- Installing fiber to each unit and mounting the ONT (Optical Network Terminal) at or inside the unit

Using existing cabling eliminates the need to install new cabling but assumes the current cables are in good enough condition to carry the signal bandwidth required, which is often not the case. Other options require installing new cables. Using wireless assumes adequate bandwidth over the wireless network, but for typical users, it may be inadequate for video.

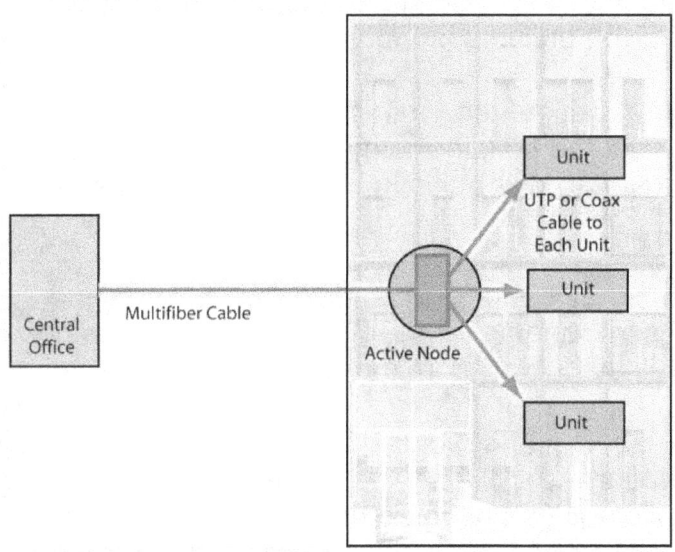

FTTH in multi-dwelling unit (MDU) using active node

Options 1 through 5 require considerable investment in electronics, space to locate them and quality uninterruptible power at the building entrance facility. The actual network architecture is influenced by the choice of electronics. ONTs are available for single users or multiple users, allowing one to distribute ONTs in a building, for example to serve all the units on one floor with copper cables. However, these multi-user ONTs are going to divide up available bandwidth among the number of units served, perhaps not a problem if the system if offering Gb/s services to the ONT, but potentially a large problem, even today but certainly in the future, if the bandwidth allocated to multiple users is much lower.

With option 6, we would generally assume a GPON or EPON system, although a point to point (P2P) system can be used. In the case of a P2P system, fiber to the unit would entail either a switch in the MDU building itself or a large fiber count cable back to the central office or nearest switch.

Assuming a GPON or EPON network, option 6, installing fiber to every unit, has several variations that can be used, and all have one big advantage: no matter how big the building and how many units, the size of the entrance facility is minimized, and no power will be required except at each individual ONT at the unit. The options start with where to place the PON splitters to optimize the cabling and installation then what kinds of cabling and hardware are needed to simplify the installation.

Options for connecting units with fiber include these architectures:
1. PON splitters can be outside the building in a service provider facility and large fiber count cables brought into the building, then broken out in premises drop cables to units. This architecture also supports a P2P (point to point, not PON) system.
2. PON splitters can be located in the entrance facility of building, minimizing the fiber count into the building, then drop cables run from that point to each unit.

3. PON splitters can be cascaded from an initial PON splitter in the entrance facility to individual splitters on each floor or area of the building, supporting units on that floor or area.
4. Theoretically, one could have a OLT (optical line terminal) installed in the building connecting to splitters distributed throughout the building or buildings. Since these units support thousands of users, dedicating a unit to one building would probably not be done except in large complexes.

The actual architecture will be influenced by the design of the MDU building and where and how it is convenient to install components for the FTTH systems. Component cost may need to be compromised to facilitate installation and reduce cost there.

MDUs come in many varieties, of course, including rows of attached units, low-rise MDUs with only a few levels of units and high-rise MDUs. The first two are more horizontally distributed while high-rise buildings can have both many vertical levels and small to large horizontal distribution depending on the height and size of the building and the size of the units.

While some older units will allow cables to be installed on the exterior of the building, that is probably not going to be allowed on more modern buildings nor on high-rise buildings. However, most buildings will have facilities for cabling even if they are so old that they only had electrical and phone services originally.

Like any FTTH system, a "greenfield" installation offers much more flexibility for designing a building that simplifies cable and hardware installation. Plans can be made to include cable conduit and/or cable trays and facilities for other network hardware. But most large MDUs have provision for cabling for services like phones and CATV, perhaps even Internet if built more recently, that offer good options for FTTH fiber installation.

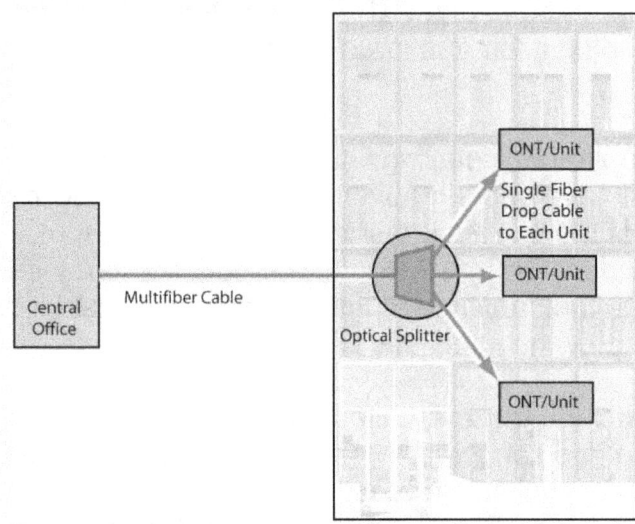

FTTH in MDU using PON architecture

The PON splitter can be located in the building entrance facility and drop cables run to each unit. This mimics most phone wiring and pathways may be available to run drop cables. It uses more fiber and/or cables but does not require mounting as much hardware around the building nor splicing and/or termination.

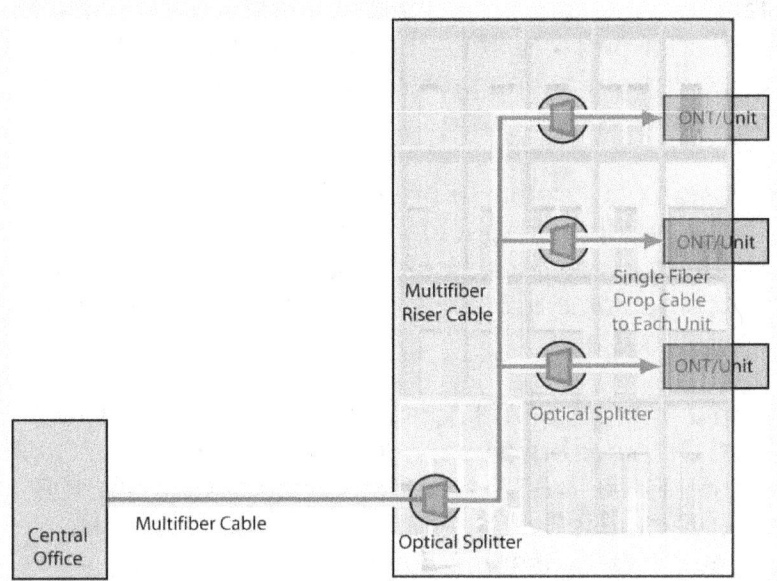

FTTH MDU with PON using cascaded splitters on each floor

One reasonable option is to use cascaded splitters. The first splitter can be located in the entrance facility with multiple fibers going out to the separate floors where a splitter is installed to serve the floor. Alternatively, the first splitter can be placed on one of the served floors. If one has 8 units per floor, a total of 4 floors can be handled on 32 split ratio system with a 4-way splitter feeding 8 way splitters on each floor. Likewise, one could use a first splitter of 8 ways to serve 4 floors. Or a 16-way first splitter would serve 4 floors.

The best option probably depends on the building and how cabling would be installed. It can also depend on the desires of the building owner or owners' association. If there is a possibility of having several service providers in the building, having an entrance facility large enough for several service providers' equipment and fibers going direct to each unit in the MDU provides the most flexibility for choice and future upgrades.

FTTH developments on distribution and drop components make MDU installations easier. Perhaps the biggest development was bend-insensitive fibers that allow the manufacture of drop cables in extremely small sizes that can be run along wall or ceiling junctions, around corners, placed inside baseboard or molding and even made with an adhesive surface that can be stuck directly on walls. Bend-insensitive fibers also allow the manufacture of small cables that allow opening at any location to break out one or more fibers for termination at that point and allow the whole cable to continue to another location.

Small boxes or closures are available that contain couplers and patch panels allowing drop cables to be terminated with prepolished/splice connectors, either fusion- or mechanical-spliced, to complete the connections.

Like any fiber or cabling installation, the actual project will be unique but be able to incorporate ideas that worked well in prior projects. If the building project is in the design stage itself, knowledgeable fiber optic designers can provide feedback that will make the installation easier, neater, and much less expensive.

The most important part of the design of a project in an existing building is a "walk through" to familiarize yourself with the building. Inspect for entrance facilities, cabling pathways and locations for equipment on every floor. Look at several units to see where it is feasible to enter the unit and place equipment. Having a familiarity with the building itself will make choosing a design much easier.

Another issue, of course, is the take rate for FTTH connections., which can affect planning as well as the ultimate cost. On older buildings, units may already have CATV or satellite connections and not be interested in FTTH, so one cannot assume a 100% take rate. The building owner can survey those living in the units to determine the take rate for planning purposes, but one also has to assume some number of future additions in doing the design. New construction may be easier, as the developer/builder may decide to make FTTH a selling feature and provide it to all the units.

Chapter 3. FTTH PONs: Passive Optical Networks

Objectives: From this chapter you should learn:
What a a passive optical network (PON)
Different types of PONs
Technical specifications for PONs
Other uses for PONs

Passive Optical Networks (PONs)

A PON or passive optical network utilizes a passive splitter that takes one input and splits it to "broadcast" signals downstream to many users. This reduces the cost of the system substantially by sharing one set of electronics and an expensive laser with up to 32 or more users. Upstream, the passive splitter acts as a combiner to connect all users to the same shared PON port. An inexpensive laser is used for the home to send signals back to the FTTH system in the central office. In the CO or head end, the OLT (optical line terminal) has a port that connects to a single fiber, transmitting data bidirectionally at different wavelengths to a splitter which connects to the ONT (optical network terminal) at multiple subscribers.

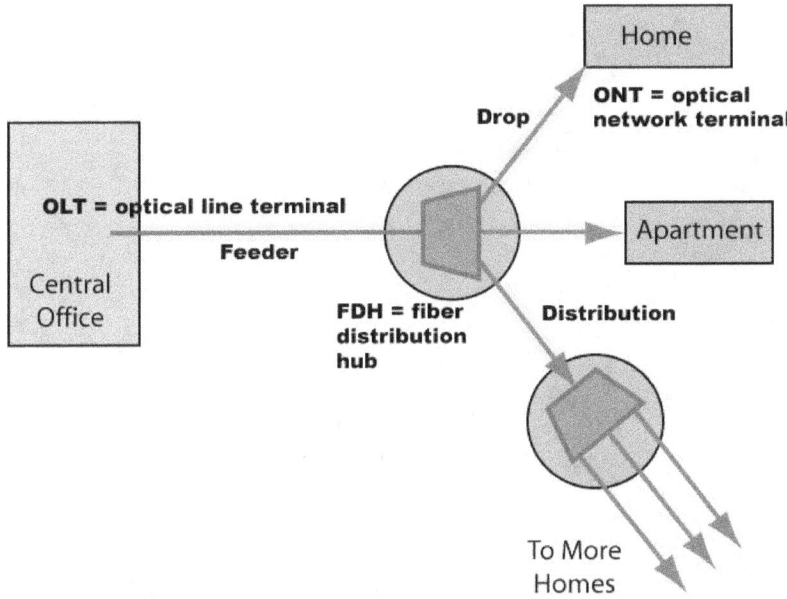

FTTX PON

Triple Play Systems

Most FTTH systems are so-called "triple play" systems offering voice (telephone), video (TV) and data (Internet access.) To provide all three services over one fiber, signals are sent bidirectionally over a single fiber using several different wavelengths of light.

Other Uses For PONs

PONs offer low cost connectivity for a large number of users with high security and relatively low management needs. Some PON suppliers have been promoting PONs as an alternative to LANs (Local Area Networks), which are especially attractive to organizations with large numbers of users. Passive Optical LANs (POLs) are less expensive than traditional structured cabling LANs but offer virtually unlimited future expansion. See the FOA Guide for more information on POLs.

APON

The first PON standard, APON, was quickly replaced by BPON because it had no provision for broadcast video and digital TV was several years away.

BPON

BPON, or broadband PON, was the first popular PON application for FTTH but most systems have been updated with GPON. BPON uses ATM as the protocol. ATM was widely used for telephone networks and the methods of transporting all data types (voice, Internet, video, etc.) are well known. BPON digital signals operate at ATM rates of 155, 622 and 1244 Mb/s.

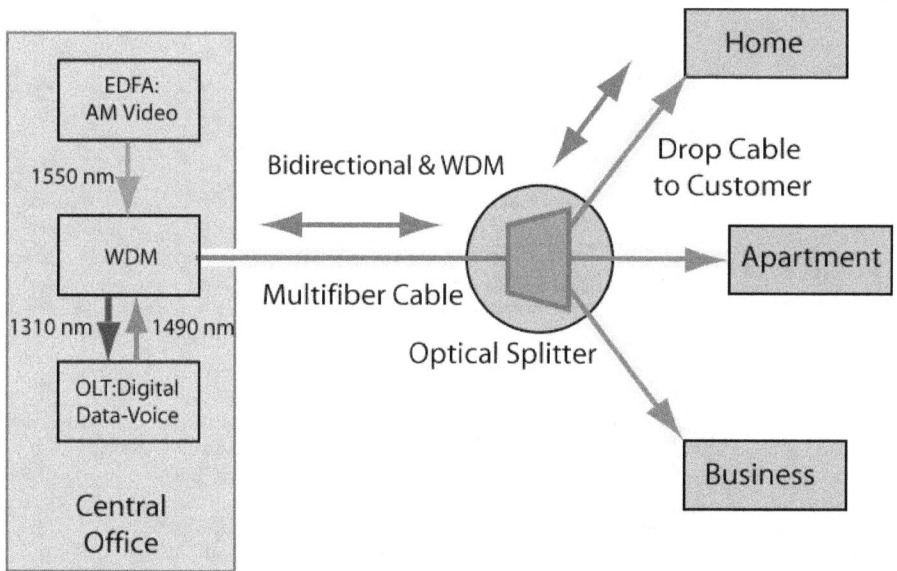

FTTH BPON architecture with analog TV

Downstream digital signals from the CO through the splitter to the home are sent at 1490 nm. This signal carries both voice and data to the home. Video on the first BPON systems used the same technology as CATV, an analog modulated signal, broadcast separately using a 1550 nm laser which may require a fiber amplifier to provide enough signal strength to overcome the loss of the optical splitter. Video could be upgraded to digital using IPTV, negating the need for the separate wavelength for video. Upstream digital signals for voice and data are sent back to the CO from the home using an inexpensive 1310 nm laser. WDM couplers separate the signals at both the home and the CO.

GPON

GPON, or gigabit capable PON, is the most popular version of FTTH PONs. GPON uses an IP-based protocol and either ATM or GEM (GPON encapsulation method) encoding. Data rates of up to 2.5 Gb/s are specified and it is very flexible in what types of traffic it carries. GPON enables "triple play" (voice-data-video) and is the basis of most planned FTTP applications in the near future. In the diagram above, one merely drops the AM Video at the CO and carries digital video over the downstream digital link.

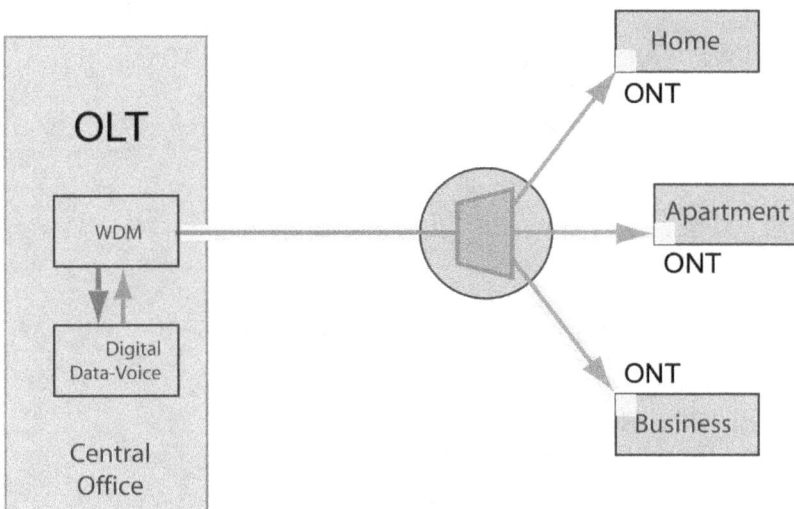

GPON adds digital IPTV to simplify the ONT

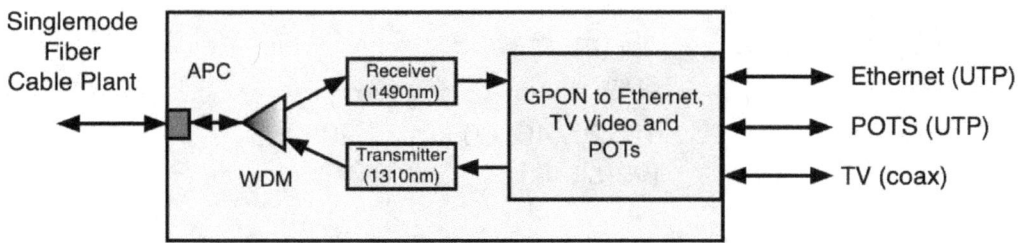

FTTH GPON ONT provides outputs for all services to subscriber

EPON

EPON or Ethernet PON is based on the IEEE 802.3 standard for Ethernet in the First Mile. It uses packet-based transmission at 1 Gb/s with 10 Gb/s under discussion. EPON is widely deployed in Asia. The system architecture is the same as GPON, but data protocols are different.

PON System Specification Summary

	BPON	GPON	EPON
Standard	ITU-T G.983	ITU-T G.984	IEEE 802.3ah (1 Gb/s) IEEE 802.3av (10Gb/s)
Downstream Bitrate	155, 622 Mb/s, 1.2 Gb/s	155, 622 Mb/s, 1.2, 2.5 Gb/s	1.25 Gb/s, 10.3 Gb/s
Upstream Bitrate	155, 622 Mb/s	155, 622 Mb/s, 1.2, 2.5 Gb/s	1.25 Gb/s, 10.3 Gb/s
Downstream Wavelength	1490, 1550	1490	1490, 1550
Upstream Wavelength	1310	1310	1310
Protocol	ATM	Ethernet over ATM/IP or TDM	Ethernet
Video	RF at 1550 or IP at 1490	RF at 1550 or IP at 1490	IP Video
Max PON Splits	32	64	16
Transmitter Power*		OLT: 0 to +6 dBm, ONT: -4 to +2 dBm	
Power Budget*	~13 dB (min) to 28 dB (max) w/32 split	~13 dB (min) to 28 dB (max) w/32 split	
Coverage	<20 km	10, 20, 40, 60 km (versions)	<20 km

* There are several versions of each type that vary so these are typical ranges.

RFOG: CATV's FTTH

CATV operators were the first broadband providers using a HFC (hybrid fiber coax) system with cable modems using RF signals. Today, some CATV operators see a need for a system to provide fiber to the home, which has lead to the development of RFOG (RF over Glass.) CATV standards have looked at PON architectures and the SCTE has proposed a standard for deploying a broadcast architecture of analog signals similar to PONs called RFoG for RF (radio frequency - i.e. FM) over Glass. RFOG is basically nothing more than an all-fiber HFC/cable modem system built with less expensive components now available thanks to the volume pricing of components used in FTTH.

It's designed to operate over a standard telco PON (passive optical network) fiber architecture with short fiber lengths and including the losses of a FTTH PON splitter.

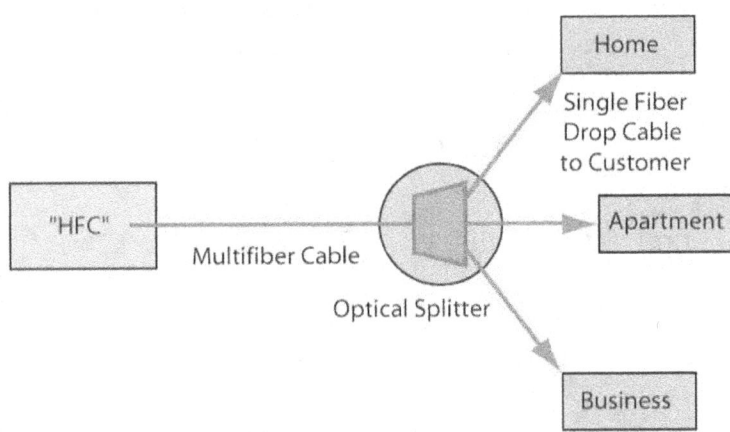

CATV PON RF over glass (RFOG) using hybrid fiber/coax electronics

There is one interesting aspect of this approach. Now telcos and CATV companies can deliver the same services over the same cable plant using totally different technologies. But that means that office or apartment building owners, developers or even whole towns that might be considering installing FTTH infrastructure themselves and leasing the fiber to a service provider can have a choice of service providers. One cable network can support either CATV or telco systems – or even someone else for that matter. That opens up a big market for private fiber optic systems.

WDM and PON

PON networks use WDM (wavelength-division multiplexing) with different wavelengths upstream and downstream to prevent signal interference. A PON sends signals in both directions on a single fiber, the downstream signals are sent at one wavelength (1490 nm) and upstream at another wavelength (1310 nm) to prevent signal interference.

Encryption

Since the OLT transmission goes to all users at once, each user has a special encryption code to decode only their data, maintaining privacy. Encryption has made PONs popular with users that want to maintain privacy like government agencies.

10 Gigabit PON Upgrades - Speed, Split & Distance

As is common with all communications networks, work on upgrading network capability and speed starts as soon as a network is introduced, and PONs are no exception. GPON has been the most widely used PON scheme for both FTTx networks and passive optical LANs (OLANs) and GPON has been upgraded to several versions with higher transmission speeds and higher power budgets to allow greater distance, higher split capability, or both.

Upgrade PON System Specification Summary

	NG-PON2	**XG-PON**	**XGS-PON**
Standard	ITU-T G.989	ITU-T G.987	ITU-T G.9807
Downstream/Upstream Bitrate	10/2.5, 10/10, 2.5/2.5 Gb/s	10/2.5, 10/10 Gb/s	10/10 Gb/s
Downstream Wavelength	~1596-1603 nm	~1575-1580	Either same as GPON if no current GPON or XG-PON for overlay
Upstream Wavelength	~1524-1544	~1260-1280	Either same as GPON if no current GPON or XG-PON for overlay
Max PON Splits	64, 128, 256	64, 128, 256	64, 128, 256+
Power Budget*	14-29 dB (min - max) up to 20-35 dB (min - max) in 4 versions with up to 15 dB differential optical path loss	14-29 dB (min - max) up to 20-35 dB (min - max) in 4 versions with up to 20 dB differential optical path loss	13-28 dB (min - max) up to 20-35 dB (min - max) in 6 versions with up to 20 or 40 dB differential optical path loss in 2 versions
Coverage	20 and 40 km versions	60 km	60 km

The assumption is that a fiber network has a lifetime of up to 40 years, so upgrades to GPON have assumed that they will use the same passive optical network architecture and fiber type (G.652 or G.657 singlemode.)

Furthermore, upgrades have been designed around coexistence with current GPON networks. By utilizing different wavelengths, it is possible to have these newer, faster networks sharing the same passive optical network as the original GPON system, allowing offering higher speeds to users while continuing to serve current users without disruption. Some commercial users can take advantage of higher speeds while typical consumers are well served by GPON. One of the big advantages of the PON upgrade standards is the ability to overlay networks. Thus a city could operate one regular GPON network for consumer FTTH use and have another, faster network operating on the same cable plant independently, offering a higher level of service and security.

PON Networks - Advantages and Disadvantages

PON networks are quite different from the typical communications network that relies on active links and switches. Here are some differences, advantages and disadvantages:
- No electronics between central office/head end (OLT) and user (ONT) means there are no electronic components that need space for mounting, power (including uninterruptible power), service or upgrades.
- Fewer electronic components and the infrastructure to support them make PONs much lower cost than P2P links - as much as 50% in capital expense and 80% in operating expense.
- Because PONs are intended to carry voice, data and video, virtually any network carrying any type of traffic can use a PON. PONs are being used for Internet service, connecting cellular sites, utility grid management, etc.
- Signals are encrypted for privacy and security, an advantage in any network.

The all-passive infrastructure means that upgrades are simpler - just change out the end electronics which will run on the same cable plant. In fact, the GPON upgrades to 10 Gb/s can run on the same cable plant simultaneously with the lower speed GPON since it uses different wavelengths. Thus, the service provider can have both low and high speed network service on one cable plant, another economic advantage of PONs. PONs share fibers to the splitters so they need fewer fibers than point-to-point networks.

Because PONs share fibers and "broadcast" signals downstream to subscribers, it requires encryption to ensure only the intended recipient can receive the messages intended for them. Thus, PONs are secure, a major advantage to organizations (governments in particular) who are concerned over security.
If there is any disadvantage to a PON network over P2P it is in the design phase where deciding on the location of splitters for an optimal system can be more time consuming than simple P2P links.

Connecting To The Internet

The PON OLT will need a connection to the Internet. Since FTTH hardware is designed for large numbers of users, perhaps an entire city, the equipment is designed to connect to very high speed Internet backbone connections. A mall system might only need connection speeds of hundreds of megabits per second, but most networks will require much faster connections, usually multiples of 1 or 10 gigabits per second.

A common question asked is how fast the Internet backbone connection should be. That's not an easy question to answer because it depends on how many users there are, what services are provided and the time of day. Video uses much more bandwidth, and more people tend to watch video in the evening, but video conferencing will be more prevalent during the day. And bandwidth requirements are increasing because of new services that use more bandwidth.

Our recommendation is that you contact PON equipment manufacturers and Internet service providers when planning a system and get their advice on equipment and estimates on costs.

Chapter 4   FTTH Network Design

Objectives: From this chapter you should learn:
Requirements for FTTH passive optical networks
How to design a FTTH cable plant
Choosing components for the cable plant
Differences between urban, suburban and rural FTTH networks

Introduction

There is really no way to generalize on the design process for fiber to the home (FTTH) networks - or any fiber optic network for that matter - since every system is unique. If you are familiar with FOA's other design materials, you know we don't give you formulas or outlines to follow. Rather than telling you how to design a FTTH network, we will illustrate some of the different network architectures, construction methods, etc. possible, then offer options that may work for your network and stimulate your design processes.

If you are new to fiber optic network design, we recommend you study the design section on the FOA Guide, read the FOA textbook *Reference Guide to Fiber Optic Network Design*, and perhaps take the Fiber Optic Network Design self-study course on Fiber U to prepare yourself for designing your own network. That will help you understand how to design and install systems most efficiently before beginning your own project. And, of course, FOA has a complete series on FTTH on the FOA Guide.

The best way to understand the options in FTTH network design is to consider several very different types of networks which differ by subscriber density, geography and technical issues which affect the design decisions that must be made.

- Urban
- Suburban
- Multi-dwelling Units
- Rural

Within each of these we will discuss design options that have been proven successful in the real world. We will look at how to design the architecture of the system as well as the design of the cable plant itself, down to the component level. After the cable plant, we'll discuss what's needed at the central office or head end. We will also look at cost implications and future upgrades. We'll also provide some cautions about fiber optic projects and FTTH in particular.

Connecting To The Internet
The first thing to consider is your FTTH network will need a connection to the Internet backbone. You need to determine the availability of high speed connections to the Internet and determine how your CO/head end will get it's connection. It may involve installing a dedicated fiber optic link from the provider's cable plant and equipment to

your planned CO. That can be a complicated project itself, requiring negotiations on not just the service to be provided, but the logistics of installing a fiber optic link including getting rights-of-way and permits. This may be a gating item for the design and installation of the entire FTTH network.

PON (Passive Optical Network)

Most FTTH networks are based on passive optical network architectures, simply because that's usually the lowest cost way to design a FTTH network as well as the lowest cost to operate. There are other architectures that may be preferable in some circumstances, and we'll discuss those too. This drawing shows the location of the hardware used in creating a typical PON network. This drawing also defines the network jargon for cables: a "feeder" cable extends from the OLT (optical line terminal) in the CO (central office) to a FDH (fiber distribution hub) where the PON (passive optical network) splitter is housed. It then connects to "distribution" cables that go out toward the subscriber location where "drop" cables will be used to connect the final link to the ONT (optical network terminal).

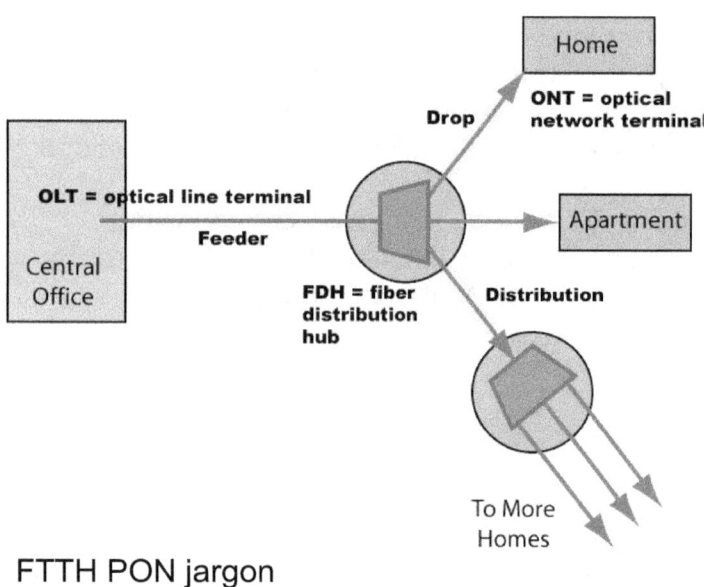

FTTH PON jargon

Active Star

An alternate to a PON is an active star network, also called a point-to-point (P2P) or "home run" system where each subscriber has a dedicated fiber and Ethernet link to the head end or central office. The main difference with a PON is the amount of fiber required for the network, especially if the service provider's switches are located at the head end. Switches can be remote, closer to the subscribers, but the switch requires power, a UPS and perhaps even heat/AC, making that option much more complex.

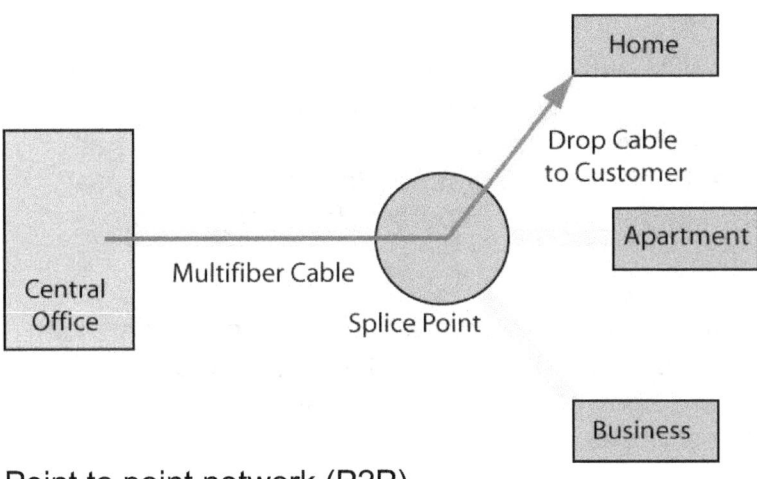

Point to point network (P2P)

We will focus on PONs but mention P2P and even some options for wireless or wireline for drops.

Gig or 10 Gig?
Standards for PONs give you an option of gigabit or 10 gigabit networks. Technically what is called gigabit (G for short) is regular GPON which offers OLT ports at 2.5 G downstream and 1.25 G upstream, shared among 32 (normal) or 64 (rare) users. Provisioning is generally for 1 Gig down and some lower value up on each OLT port.

If you don't understand the stochastic nature of networks, you might assume that 32 users with 1Gb/s of bandwidth have only about 30 Mb/s (1 Gb/s = 1000 Mb/s, divide by 32 = 31.25 Mb/s). But that's not realistic. The full 1 G of bandwidth is available to all users who only use it a fraction of the time, so their data is transferred at 1 G speeds. Average usage for most FTTH networks averages only a few Mb/s, so 1 G PONs are very effective.

However, PON head end electronics generally allow to program how much bandwidth is available to any user, a technical solution to a commercial issue, how to charge more for higher bandwidth connections.

10 G PONs are available and like all electronics, costs keep coming down, so they are being considered for many networks. It is doubtful that any FTTH network aimed at consumer subscribers needs 10 G, but some business customers might. The GPON designers were clever, however, making 10 G use different wavelengths than 1 G, so if you build a 1 G GPON network, you can upgrade at any time - say to accommodate a network expansion aimed at businesses - and run both networks simultaneously over the same cable plant. Generally, we recommend building networks at 1 G to take advantage of the lower cost electronics but knowing that upgrades can be made simply using the same cable plant.

Network Architectures

PONs have options on architectures that can affect cost and ease installation. One of the first decisions the designer needs to make is where to locate splitters as that will affect other hardware decisions like how many fibers should cables have and what types of hardware to use. To make that decision, one first needs to understand the distribution of subscribers as location and density are important for designing an efficient system.

PONs work on the principle that splitters allow one central port to communicate with 32, 64, 128 or more users over a single fiber to the splitter and then a single fiber to each user. Typical PON architectures are shown like the drawing above where the splitter is at some location near the subscribers.

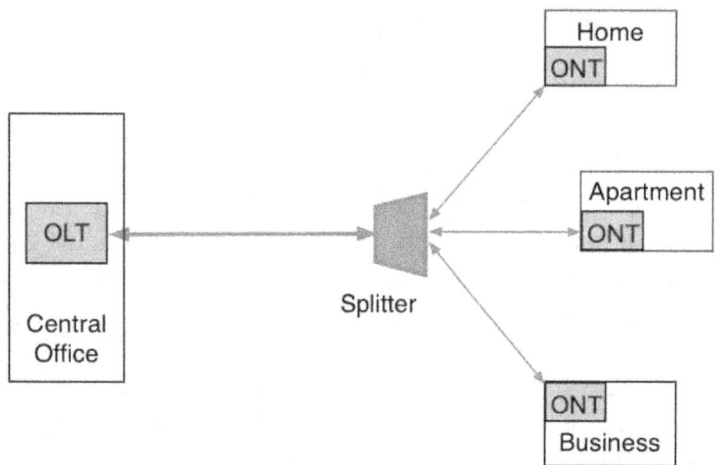

There is a lot of flexibility in the location of splitters. For example, some dense urban or suburban networks move the splitter into the Central Office (CO - a traditional telecom term) or Head End (the CATV term) and run a fiber to every user.

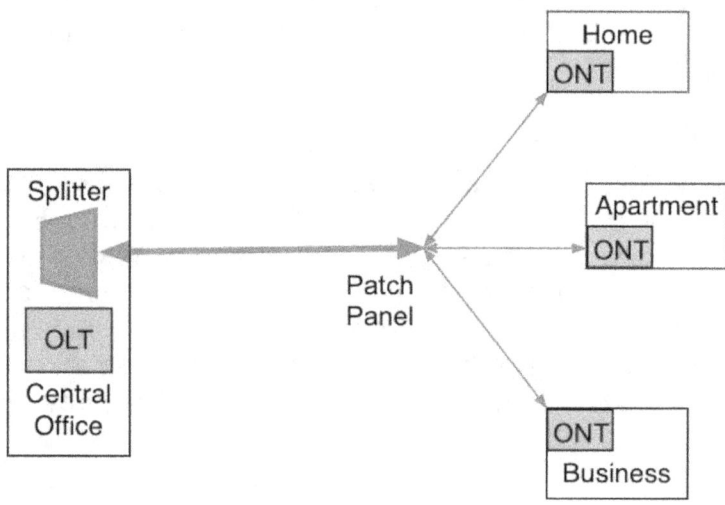

While this option requires more fiber, large fiber count cables are readily available and fiber cost is low, so the incremental cost to use more fibers in a cable is reasonable. If the splitter is in the CO, the OLT can be utilized more efficiently since each port can support 32 or more users from any location in the service area. If splitters are moved closer to the users, some ports must be left open for future expansion, meaning that OLT port will support fewer than the maximum number of users - about 24 ports of 32 available being used seems average, with 8 ports being open for future subscribers.

With a central splitter and fiber to each user configuration, there is flexibility to use each OLT port more efficiently, adding new OLT ports only when needed, and when every customer has a dedicated fiber, the CO can even support several ISPs (Internet Service Providers) by simply patching the user to their chosen ISP.

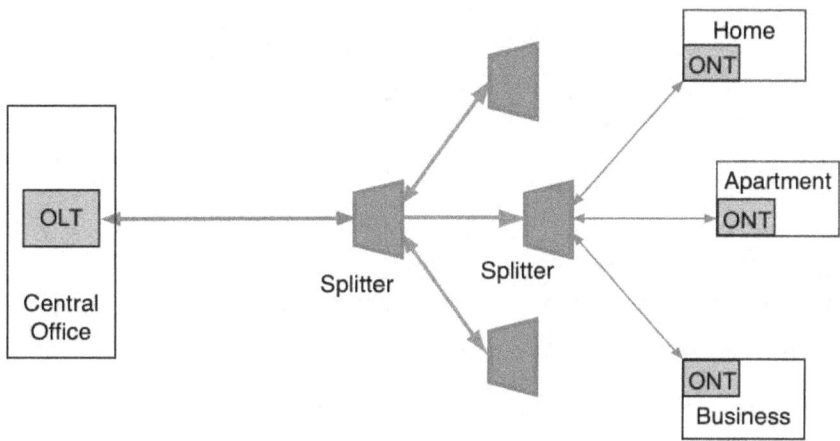

Where subscriber density is lower, it's common to cascade splitters where splitters with fewer splits are connected to other splitters in series. This can save significant amounts of fiber. Cascaded splitters are useful for areas like the suburbs or rural areas where subscribers are spread out but often in clusters. An example is to use a 4 or 8 port splitter to serve a street in the suburbs or a small cluster of homes in a rural area or to connect multiple users in a multi-dwelling unit (apartment or condo building - see the section on MDUs below.)

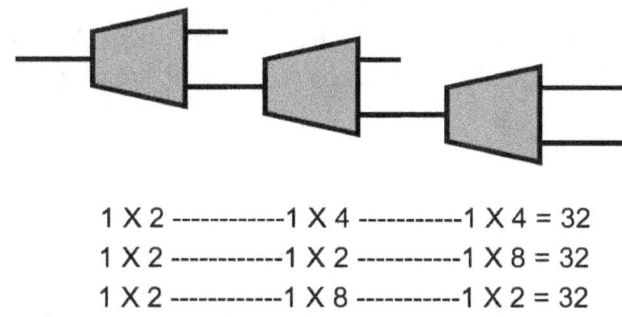

1 X 2 -------------1 X 4 ------------1 X 4 = 32
1 X 2 -------------1 X 2 ------------1 X 8 = 32
1 X 2 ------------1 X 8 ------------1 X 2 = 32

Cascaded FTTH splitters

PONs are standardized on splits of 32 or 64 per OLT port, and the economics of the electronics depends on the efficient use of the OLT port. But subscriber take rates are not 100%, so one needs to leave spare ports for drops to future subscribers. Some systems limit subscribers to 20-24 per port to allow new subscribers. Others have built networks around a fiber from the head end to every potential user and leaving the fibers to non-subscribers dark. The central office houses all the splitters which can be fully populated, optimizing the electronic ports. New subscribers already have a dark fiber and just need plugging into a splitter at the head end. New OLT ports are added as subscribers fill up ports already in use.

Much of the design time is likely to be spent deciding where to place splitters to optimize the cable plant. In dense urban areas, there may be locations where subscriber density is high enough to justify using a single 32 port splitter. In less dense areas, it will probably be more efficient to cascade splitters to equal the 32 splits. Splitters come in binary ratios (2, 4, 8, 16, 32) and can be cascaded in any sequence as log as the multiplied split ratios are no more than 32, e.g., 2+16, 4+8, 2+4+4, etc.

Choosing splitter locations can be challenging, but generally it is done where one finds groups of subscribers in close proximity, so the length of drop cables is shortest. In urban or suburban areas, one can look at the number of residents in a building or the number of homes on a street. In dense population areas, a pedestal or underground fiber distribution hub (FDH) containing splitters may be placed in a neighborhood and drops run from the FDH to buildings from there.

If a building has many residents, a smaller FDH with splitters may be placed inside the building with individual fibers run to each subscriber. A large FDH is not needed for splitters. Splice closures often have provision for splitters, so a backbone or distribution cable can be split out to drop cables for subscribers in the closure. That closure can be in a manhole or handhole if the cable plant is underground or suspended with aerial cable.

PON Loss Budgets

PONs are designed around a certain split ratio for OLTs and have a range of power at the receiver that must be met for the network to function. For example, GPON has a Power Budget range of 13 dB (min) to 28 dB (max) w/32 split. When calculating the loss budget, one needs to ensure the loss is at least 13 dB but less than 28 dB, usually less the acceptable loss margin for the link. The minimum of 13 dB is usually not going to be a problem since splitters add large amounts of loss.

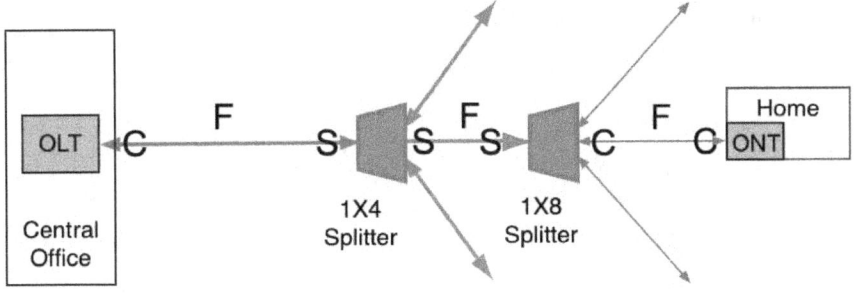

F=fiber, S=splice, C=connection, X=splitter

PON network diagram with loss budget components noted.

PON networks with splitters require calculating a loss budget like any other network. Besides the losses from fiber length, splices and connections, one must add in the loss of the splitters. Each split of a factor of 2 loses 3 dB and in addition there are losses due to splitter inefficiency, so losses can be quite high. In the example above, the splitters are cascaded so one would have a 1X4 with 7 dB loss and a 1X8 with 11 dB loss to include in the loss budget for a total splitter loss of 18 dB. The loss of the remaining components in the cable plant (fiber, splices and connectors) would be added to the splitter loss.

PON Splitter Ratios And Losses

Splitter Ratio	1:2	1:4	1:8	1:16	1:32	1:64	1:128
Ideal Loss / Port (dB)	3	6	9	12	15	18	21
Excess Loss (dB, max)	1	1	2	3	3	3	3
Loss (dB)	4	7	11	15	18	21	24

After calculating a loss budget, it needs to be compared to the power budget of the PON version or link (if using P2P architecture) chosen for the network. The loss budget for the cable plant must be less than the power budget of the link or network for the design to work.

Rural Architecture Options

Rural areas are characterized by low subscriber density and long distances, not the conditions PONs were designed for. There is a "long reach" GPON version with a capability of 64 users and 60 km that can work in some applications. Another option sometimes considered is not using splitters but taps, special splitters at drops that are not symmetrical - multiple equal outputs - but split off a small portion of the signal in the fiber, like 10% and pass 90% along to the next drop.

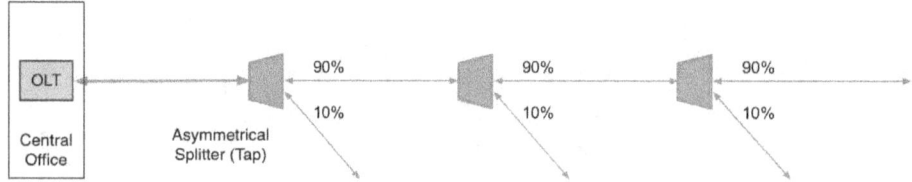

FTTH network with fiber optic taps

The problem with the tap architecture is the inefficiency of tap splitters, The excess loss in taps can be as much as 1 dB per tap. That excess loss adds up fast, rapidly cutting the length the network can reach (e.g. 1 dB = ~2.5 km of fiber). FOA has done an analysis of the use of taps in rural FTTH which can be used as a model for analyzing the use of taps in any FTTH network.

Another option which has been developed for low subscriber density like rural areas uses is a remote OLT with only a few ports. OLTs designed for CO use generally have options for many OLT ports because they are intended for applications with large numbers of users - hundreds of thousands in a dense city. But rural areas may only have a few subscribers in a small town or scattered along country roads. Large numbers of ports are not needed. What is needed is an architecture that allows a service provider to connect users spread out over large areas.

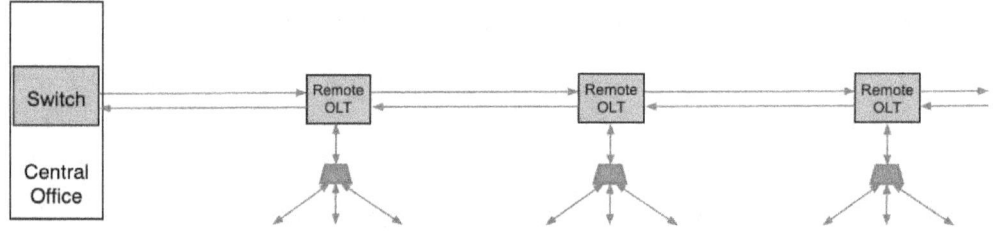

FTTH network with remote OLTs

The remote OLT option allows creating a "head end" periodically along a rural road where users are grouped. The Remote OLTs can take advantage of the fibers already installed along many roads and since some allow "daisy-chaining," the use of two fibers along the route is all that's needed. They are in small enclosures similar to CATV amplifiers and can be mounted on poles or suspended from messenger wires.

As mentioned above, the long reach version of GPON is another option for rural areas. It can reach 60 km with up to 64 users per port, but in countries like the US, 60 km is short compared to the distances in some rural areas. And it will use much more fiber than the remote OLT architecture.

FTTH Fiber Optic Cable Plant Design

When FTTH using PONs first began being installed around 2005-7, it was considered a extension of regular telephone systems, where subscribers were being connected to a telephone system replacing copper wires. Cabinets or pedestals containing the PON couplers were placed near a group of subscribers. Cables were pulled between the cabinet and the central office containing the PON system electronics and spliced on each end by the usual outside plant (OSP) installation crews and were tested as was normally done with telecom fiber optic networks.

On the subscriber end, drop cables were placed to the home and connected either by splicing or installing connectors (usually APC connectors to prevent reflectance problems). Drop cables could be installed aerially, underground or buried. Installing the cables through customer's yards created a problem as it is time consuming and disruptive to the customer. Simple trenching was sometimes dropped in favor of directional boring, an expensive process. Connectors were installed either by fusion splicing on pigtails or using splice-on connectors.

After the cable plant was installed, the optical network terminal (ONT) was installed at the home. Some systems installed ONTs on the outside of the house, some inside garages, some inside the home. Some home builders built new homes with provision for the ONT inside the home and installed cabling and power to the same area to create a home prepared for broadband. See the examples below.

After the ONT was installed and tested, it was necessary to complete the installation by connecting the customers phones, TVs, and computers. In all, three or four groups of installers of different experience and skill levels were needed to install a FTTH customer.

Systems Evolving With Experience

After some experience with the systems, methods were tried to simplify the process and cut costs. A big breakthrough came with the development of prefabricated cabling systems (sometimes call pre-terminated cabling) that eliminated the need for most of the splicing. Cables with weatherproof connectors were purchased already made to the lengths needed and pedestals were factory made with connectors for the drop to the home and a cable ready to splice onto the cable installed from the central office.

The prefab drop cables could be run aerially, even lashed to current telephone wires. They were also small enough they could be pulled through small PVC conduit often installed to home in new construction. Most of the systems use multi-connector cables near the homes being connected so homes can be connected during the first install or later when more customers decide to take the service.

Aerial installation in Santa Monica, CA, using prefab drop cabling.

If the cable is underground, it will usually be pulled through conduit from connection to the distribution cable or the splitter to the home. Above is a preterminated system has two home drops connected to the distribution cable.

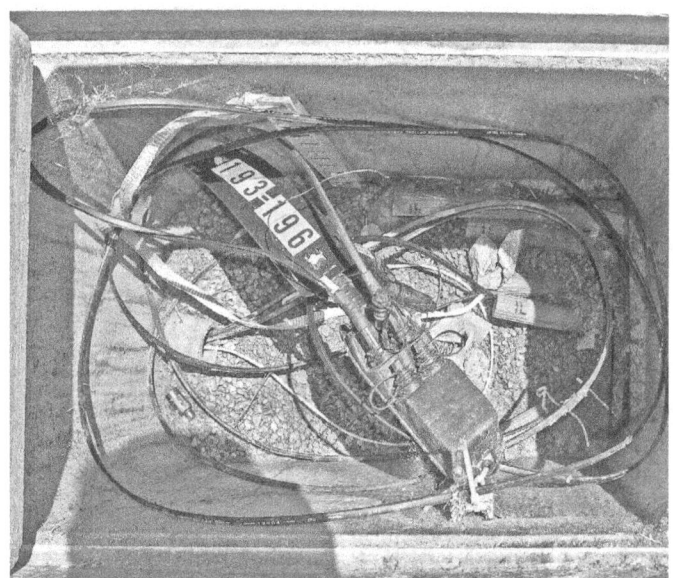
Underground installation of prefab cable system in Long Beach, CA.

In urban areas, installing cables can be done by trenching but it can also be very disruptive. If a city is doing installation or repair on any underground utility, it is advisable to install some fiber ducts for potential future use, eliminating the need for additional construction in the future. This is now polity in many areas; the US FCC refers to it as "Dig Once."

Directional boring is a construction technique used to reduce the disruption of construction and it is effective as long as the contractor is experienced using the technique and other underground utilities are carefully located before boring. Many gas lines, sewers and water mains have been punctured by contractors, so careful screening of contractors doing directional boring is important.

Directional boring in a downtown street

Other systems used microduct installation which requires little or no digging to install underground or under a road. Microducts are designed to accommodate small microcables that are installed by "blowing in" the cables, a technique used in many systems.

Microduct installation using a sawed groove on a paved street.

The splitter can be housed in a central office, a pedestal or even a splice closure used for drops in the neighborhood near the homes served. The advantage of PONs is that the pedestal or splice closure is passive - it does not require any power as would a switch or node for fiber to the curb.

FTTH pedestal

Above is a typical urban pedestal that has connections to the CO, splitters, and fibers out to each home in a sealed enclosure. Pedestals like this can be specified and purchased ready to install; installation involves only splicing the distribution cable to a short cable provided as part of the pedestal. In the photo below, a tech is splicing drop cables in a closure that has provision for a splitter in the splice tray.

FTTH closure w splitter from a rural aerial installation

A network interface device called an ONT (optical network terminal) containing a fiber optic transceiver will be installed at the subscriber. Some are installed on the outside of the house, others are indoors. Some houses are now being built with cabinets in the house for connecting to the FTTH fiber and then distributing phone, TV and Internet connections throughout the house over state-of-the-art cabling. The incoming cable needs to be terminated at the house, tested, connected to the interface and the service tested.

Basic Component Selections

Two components of FTTH cable plants are almost universal. Every network should use standard singlemode fiber, G.652, or its bend-insensitive equivalent G.657, in all the cabling. This is regular SM fiber and is appropriate for current PONs as well as upgrades to 10G PONs in the future. Connectors are generally SC-APC, the SC connector with an angled physical contact ferrule to reduce reflectance problems in the short cables and splitters. Check equipment requirements to see if SC-APC connectors are specified or if SC-PC is required for the transceivers. Beyond these basic choices, cable designs and other hardware will be chosen for the environment of the cable plant.

New Cable Types And Hardware For Subscriber Drops

Several new cable types were developed for use in FTTH. Until FTTH, most single fiber cables were complicated structures with tight buffered fibers and aramid fiber strength members inside plastic jackets, usually 3 mm in diameter or sometimes smaller. While these cables were adequate for factory termination into prefabricated assemblies, they were not ideal for field termination or use inside buildings. With the advent of bend-insensitive fibers that required less protection, a new type of drop cable was developed that molded a bend-insensitive fiber inside a small plastic structure surrounded by metal or aramid fiber strength members. This design could also be made as a "figure 8" cable with a messenger for support in aerial installation. Here are some photos of this type of cable.

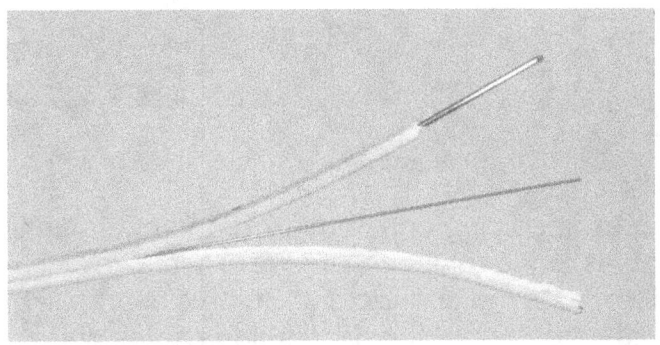

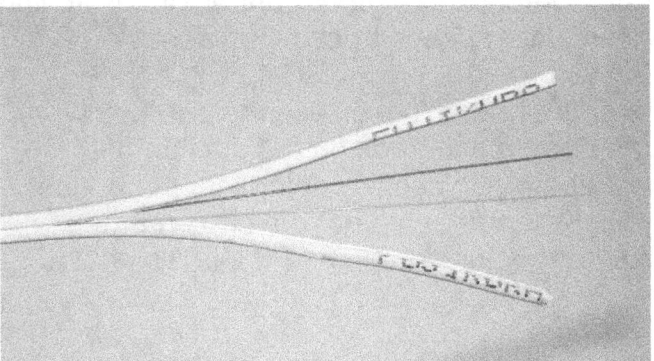

FTTH Drop Cable, 1 fiber (top, showing steel strength member), 2 fiber (bottom)

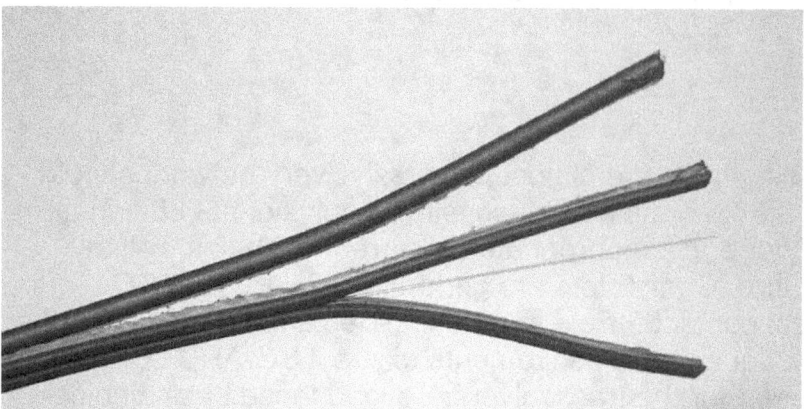
FTTH Drop Cable, 1 fiber with messenger for aerial support

To work with these cables, special fiber closures were developed that are more convenient for field installation.

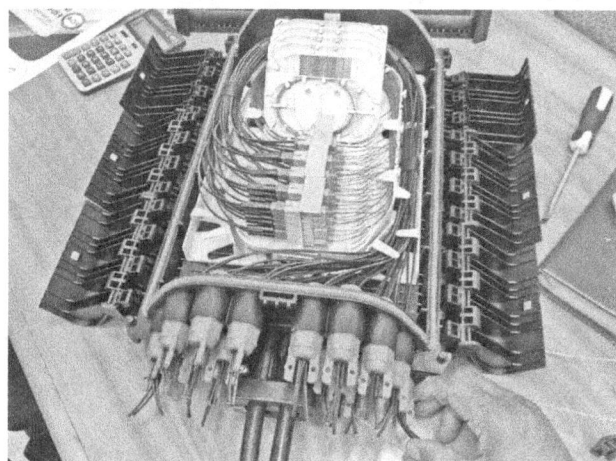
FTTH Drop Cable Closure

This closure has entries for distribution cables, including one coming in and one going out - continuing on to another closure for cables using midspan access. There are multiple outputs for drop cables which are terminated in connectors. Some closures like this one have provision for splicing on pigtails to terminate the distribution cables while others are designed for direct termination using prepolished-splice connectors. Patching with connectors in a re-enterable closure allows adding new drops when needed.

Early ONTs being installed on the outside of the house looked like the one below. Most networks have moved to just installing a demarcation box on the outside of the house where the drop cable is connected to a fiber optic cable running into the house. This connection allows for a test point for techs where they do not have to enter the house to see if signal is being received at the house. Inside, the FTTH ONT is now a small box like a cable or DSL modem.

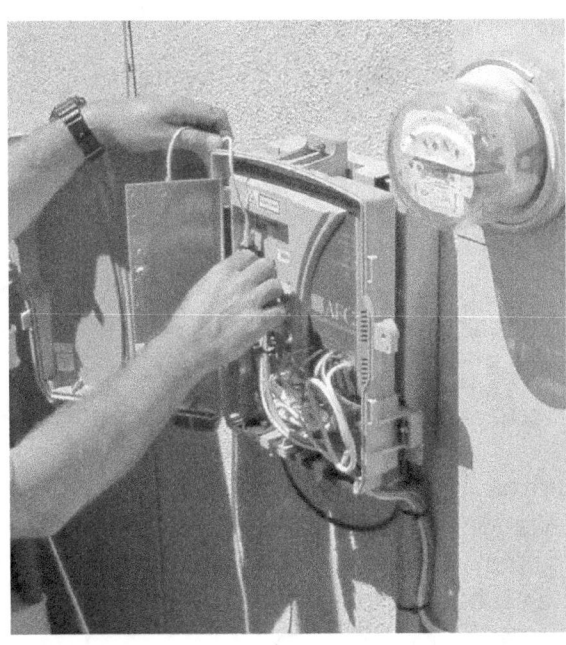

A house with premises cabling has this cabinet inside the home for distributing services from the FTTH connection. Some builders now include equipment boxes like this in new homes.

ONT in a home wired with structured cabling throughout the home

Design For Geography
The best way to understand the options in FTTH network design is to consider several very different types of networks which differ by subscriber density, geography and technical issues which affect the design decisions that must be made.
- Urban
- Suburban
- Multi-dwelling Units
- Rural

Urban FTTH Networks

Urban FTTH networks feature high subscriber density requiring less fiber because the distances are shorter and more splitters and head-end electronics because the subscribers are more numerous. In dense urban areas, many if not most users are going to be in MDUs, so the OSP design and installation can be simpler - just getting fiber to the building - but inside the premises it may perhaps be more complicated - how to run fiber inside the building to each subscriber. See the section on MDUs below.

Inside a city, the most complicated part of designing a network is accommodating the cables needed to get from the CO/Head End to the building where the users are located. Cities often have congested conduits with too many cables already, so running more cables can be a problem. Techniques exist to add cables to crowded ducts, sometimes removing fiber ducts that have only one fiber and replacing them with microducts with multiple microcables or using fabric ducts that can double or triple the number of cables a conduit can accommodate. Rather than digging up streets, systems can be built using microtrenching with minimal impact. Techniques also exist to remove the core of CATV hardline coax and use the jacket as a fiber duct and use robots to install cables in sewers. Fiber optic companies can be quite inventive.

Most cities have the majority of cables underground, but many also have aerial cables in alleys. If possible, running FTTH cables in alleys behind buildings and using aerial drop closures will greatly reduce the cost of building an urban FTTH network.

But before deciding how to install more cables, inventory what exists already. Many cities have fiber optic cable plants installed for city communications, security, traffic systems, citywide WiFi, etc. and may have spare fibers. If no spare fibers exist, it may be possible to use wavelength division multiplexing to get more links. Likewise, other service providers including electrical utilities may have fiber or spare conduits that can be used, and their interest may be higher if they can benefit from the new network.

Another problem in cities is finding space for fiber hubs where drop cables connect to the distribution cables. They can take up lots of space and be a problem in cities where sidewalk space is at a premium and underground utilities crowd the areas under streets and sidewalks. Putting hubs inside buildings may be much easier that doing major construction outdoors.

FTTH In Multi-Dwelling Units (MDUs)

Multi-dwelling units (MDUs in jargon) are sometimes handled like FTTC, where fiber is brought into the building and individual units are connected over copper cables, either twisted pair from phones or computer networks or coax from CATV or satellite. A standard for Ethernet over CATV Coax (MOCA) is often used to make this connection if the telephone wires are inadequate but be careful as those cables may not be owned by the building owner but the service provider.

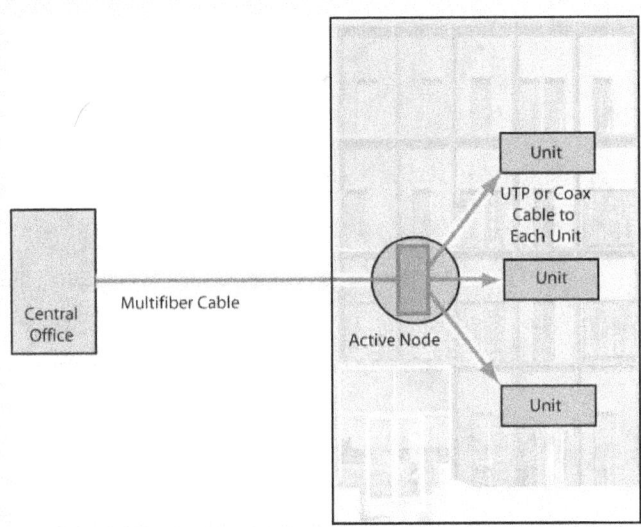

MDU with active node like FTTC

But MDUs are ideal for FTTH (that is to each unit in the building) since there are many users in a very small space and fiber lengths are short. Besides using less fiber, MDUs generally require less time per drop to install. One issue is where to place PON splitters. If it is a small building, the splitter(s) can be installed at the entry facility and individual cables run to each unit. In larger buildings, splitters can be cascaded, and a splitter placed on each floor (if space permits) and short cables run to each unit.

Each building should have some space for the fiber to enter the building, have a rack or box for splitters and connecting to cables to run to each user. Since the PON network is passive, it is not necessary to have power at this location, just some space and room to work on the hardware connecting or moving users.

A major problem in older buildings has been finding places to run cables to each subscriber, but new types of bend-insensitive fiber and the special small drop cables shown above make it easy to route fibers along walls or place in stick-on raceways on the walls. Cables installed in buildings should be done neatly. Residents and owners are likely to complain if the cabling is done badly.

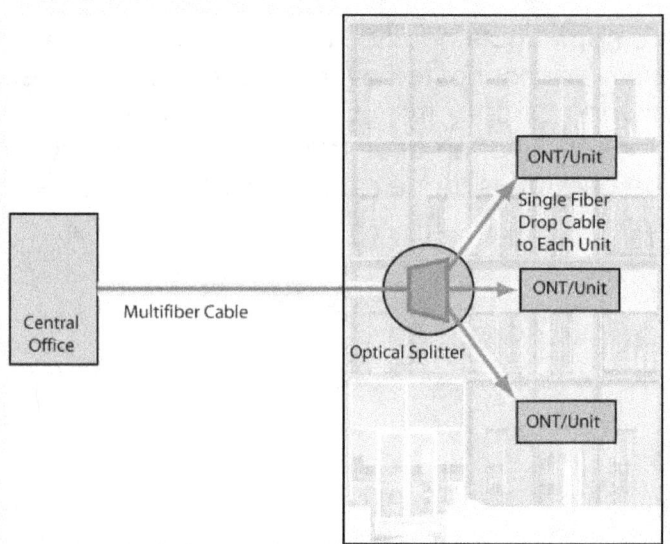

MDU with fiber to every unit

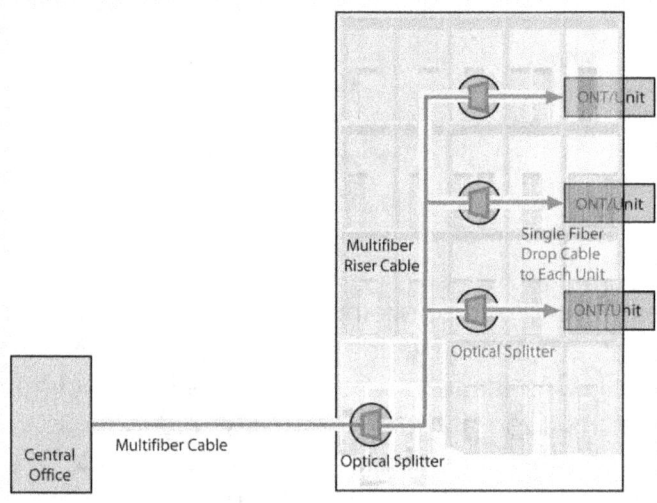

MDU with cascaded splitters

Suburban FTTH Networks

Suburban networks are less dense and installing new cables can be much easier. Some areas have aerial infrastructure which makes installation much easier, since aerial installation is always easier and less expensive and special FTTH closures can allow placing fiber hubs with splitters on poles or suspended from messenger wires. Aerial drops are easier to individual homes or buildings also. If the cables must be installed underground, which is becoming more common, and conduit space is not available, microtrenching along the curb can simplify the installation. Handholes will need to be located where splitters and drop closures are located.

Microtrenching can hide cables very effectively in urban or suburban neighborhoods

Underground drop cables to each home can be a problem since the cable or ducts needs to go from a curbside handhole through the customer's lawn. Running the cables alongside the driveway alleviates most of the problem in many drops. If the drop cable is buried in the lawn, there can be a problem in the future with the owner digging up the cable because they forgot where it was located, or a landscaper or installer of invisible dog fences starts digging. As close to the driveway as possible is safer.

Directional boring is also possible but can be extremely expensive and has the danger of hitting current utilities and causing damage. Not all - probably few - owners know where their underground utilities are located and using locating equipment is time consuming and expensive. Gas lines are especially dangerous and can cause fires and explosions.

From the outdoor demarcation box, a fiber needs to be run to a location inside the home to the location of the FTTH ONT. In any indoor installation, it is important that the work be done neatly. A survey of the home to locate where other cables (telephone and CATV or satellite) enter the building and are run indoors can help locate the easiest path for the FTTH fiber cable. As with MDUs, the resident/owner will expect the cables to be installed in a neat and workmanlike manner.

Rural FTTH
Rural FTTH is going to be more expensive no matter what you do. Service providers have used wireless drops to avoid running long fibers to each subscriber, but the limited bandwidth, cost of the equipment on a pole and at the subscriber (not to mention updates) and the lifetime of powering it may prove to be no bargain compared to running a fiber optic drop cable to the subscriber. Most rural networks are aerial which makes the cost much lower than underground. If the utility poles already have low voltage cables, generally aging telephone wires, the FTTH fiber distribution cable can

probably be lashed to the same messenger, by far the least expensive way to install fiber.

Splicing FTTH backbone cable in the Mojave Desert

The big problem with rural FTTH is distance. That means longer cables and more cable and installation costs. It also means that many networks will exceed the length maximum of GPON while having fewer subscribers than the number allowed by splitting. The solution in many cases may be using the remote OLT architecture shown above. Cables can be run alongside rural roads and the remote OLTs installed in areas where there are multiple users within the reach of the OLT.

Electrical utilities and coops have rights of way and transmission lines into rural areas that are ideal for FTTH. Many utilities already have fibers installed, often in optical power ground wire, but sometimes have only a few fibers available. WDM (wavelength division multiplexing can expand the fibers capacity if permitted. When you already have towers or poles, installing fiber is easy with several options. Cable can be lashed to messengers even overlashing to current fiber or copper cables. ADSS (all-dielectric self-supporting) fiber optic cable can be installed on poles or towers without a messenger for long spans. There are even methods that use lightweight cables that can be wrapped around current-carrying conductors.

Rural telephone companies also have rights of way and cable infrastructure. Many already run fiber backbones. For them, the biggest obstacle is usually financing and fortunately there are programs to provide grants and loans for rural broadband.

Cable cost is higher for the longer distances, but fiber optic cable is very inexpensive, especially when compared to installation cost. The cost of fiber itself is only part of the cable cost; the materials used in the cable predominate for cables with fewer than 24 fibers, so buying or installing cables with fewer than 24 fibers is generally not cost effective. And installing more fibers means that spare fibers can be leased to other carriers - municipal/county/state agencies, telecom companies, wireless service providers, electrical utilities, etc. to defray the cost.

Designing The FTTH Central Office or Head End
Whether you call it a central office or head end probably depends on whether your background is telco or CATV. IT types might call it the equipment room or data center. But the location of the electronics for a FTTH network is another design decision that requires careful consideration and planning.

The location should generally be as central as possible to simplify and perhaps minimize cable lengths. Space is not a big problem as FTTH PON OLTs are not large because each OLT port serves as many as 32 or more users. But the design must accommodate cables and electronics and allow for future expansion.

Head end electronics for approximately 5,000 subscribers

The head end is also an entrance facility, with incoming fiber optic cables from the Internet backbone connection and outgoing cables from the OLT to the FTTH cable plant. Sufficient patch panel space is required for both incoming and outgoing cables. Hardware needed includes cable trays and management as well as rack space for the patch panels.

There will have to be rack or wall space for the interface to the Internet service provider. This is essentially a router with incoming signals at high speeds. The router connects to the OLT to handle the Internet connections. This service requires backup power just like the OLTs and other associated equipment.

The OLT connects the Internet to the users by converting signals to PON protocols and connecting users. The OLT does lots more also, including managing users and encrypting signals for subscriber privacy. The OLT requires a trained operator, at least part time, to manage subscriber connections and other regular service, so plan on having an operator attend training by the equipment provider.

Since the subscribers require 24/7/365 operation, the entire electronics package needs conditioned power and an uninterruptible power supply (UPS) that can keep it going for at least 8 hours in a power outage. Many FTTH systems also provide – or at least recommend – a UPS at the subscriber's site to also provide nonstop operation.

Operating Personnel
The OLT will also require some management to add/remove users or change operating parameters such as the speed of a user's connection. This will require at least a part-time technician with some training from the supplier of the FTTH equipment and perhaps the router that connects to the backbone also.

There will also be a need for customer service personnel to assist users with problems. They need to be able to troubleshoot ONT problems and connections to the users equipment.

Since users will be upset if there are service outages or delays in making changes, the personnel involved in operating the system need to be available at least during the normal workday.

Before you Start, More Things To Remember
Here are some things FOA has learned from past FTTH projects:

Uniqueness: Like most fiber optic networks, every FTTx installation is unique. It must be designed for the location it is to serve and choices on components and installation methods should be optimized for the system. Construction and installation methods may include every type of OSP installation. Suppliers familiar with FTTx can advise customers on what others have done to make installations simpler, easier and less expensive. Most systems prefer to use as many factory-made components as possible as they are generally less expensive than doing the same work in the field. New installation methods should be considered as well to reduce costs.

Consultants: Be wary of consultants. Consultants can be extremely valuable in designing a FTTH system, as long as they have relevant experience, are up to date on new components and techniques and are highly recommended by previous clients. Unfortunately we have seen problems with consultants, including over-designed networks with costs much higher than necessary, installation practices recommended that were unnecessary or ignore newer technology, systems designed around components that were higher performance (and price) than necessary, and in one case a consultant took the clients payment, went away for a year and came back with an admission that they could not design the network (but they kept the consulting fees.)

What Fiber Do You Already Have?   Before you design or install a new fiber optic cable plant, inventory the fiber you have already and/or negotiate to lease fiber where others have cables with dark (unused) fibers. Also talk to other organizations who may need communications to see if they want to share costs or lease dark fibers or communications links from you. Cities, counties, and states need fiber. Utilities need fiber. Fire and life safety organizations need fiber. Traffic departments need fiber. Cellular companies really need a lot of fiber.

What Other Services Can Share The Fiber?   Consider what other services than FTTH you can carry on your fiber optic cable plant - cellular backhaul, traffic systems, security/surveillance systems, leased fiber, etc. to generate additional revenue. A few years ago a large American city sent out a RFP (request for proposal) for an urban FTTH network. The document dealt strictly with FTTH to connect the city's citizens with fiber and ignored all the other services the city had that already used or needed fiber - city communications, security/video surveillance, intelligent traffic management, public transportation communications, wireless networks (small cells and 5G), utility communications, etc., etc., etc.

Dig Smart -Dig Once: This same document also covered the difficulty of urban installation - digging up streets already filled with underground utilities, limited space for pedestals, few options for aerial cable and other issues that are typical problems for urban fiber installation. No mention of "Dig Once" to make future installations easier. Share fibers. Use spare fibers. Use additional wavelengths in current fiber. Consider all the alternatives. Plan ahead - future proof is a myth, but one can make certain decisions that will make the future easier.   If you are considering using FTTH design software, ask to talk to customers who have used it. Determine what you need to know first in order to use it, e.g. GIS data on every utility pole, manhole or handhole, subscriber location, etc. and how much training it takes to become proficient. Will you use your personnel or hire outsiders, and how do you evaluate them?

Cost Savings: Fiber optic cable and components are not expensive, but labor is. Saving money on components may look good in first analysis, but more savings will come from optimized designs and efficient installation practices. More experienced contractors are more efficient and may save costs by their speed and efficiency. And design for the future - if you dig a trench for anything, not just fiber but any underground utility, bury several fiber ducts for future use, install cables with more fibers than you need - lots more - fiber is cheap, installation is expensive. The program is called "Dig Once."

Take Rates Are Important: "Take rates" for new FTTH networks vary from low to high, depending on the subscriber's satisfaction with the current ISP (Internet service provider.) When Google Fiber started in Kansas City, the take rate was high because the current service was bad, but in later cities when the local ISPs knew they were coming and improved their service and/or lowered their prices, the take rate was lower. Competition tends to drive take rates and take rates determine the economics of the system, Know your competition. Offering gigabit services are often the top selling point of FTTH. Every GPON network is a gigabit network, but subscribers can opt for slower speeds at lower costs.

Chapter 5. Fiber To The Home Installation

Objectives: From this chapter you should learn:
How FTTH fiber installations are similar to and/or different from other installations
What is involved in customer premises FTTH installations

Introduction To Installation

There is probably no way to generalize on the installation process for FTTx since every system is unique and, in some cases, every subscriber is different. Rather than telling you how to install FTTx here, we will try to illustrate some of the ways that others have installed their systems and offer advice on how to install systems most efficiently.

Instead of duplicating information elsewhere in the FOA Guide, which has a long section on fiber optic construction and outside plant installation, we will focus on FTTH specific topics and link you to some FOA online materials that cover relevant topics. We also recommend you read the FOA Guide pages on FTTH and especially the page on FTTH Network Design before starting on this page.

Like most fiber optic networks, every FTTx installation is unique. It must be designed for the location it is to serve and choices on components and installation methods should be optimized for the system. Installation methods may include every type of OSP installation. Suppliers familiar with FTTx can advise customers on what other systems have done to make installations simpler, easier and inexpensive. Most systems prefer to use as many factory-made components as possible as they are generally less expensive than doing the same work in the field. New installation methods should be considered as well to reduce costs.

Contractors need to be well trained (and preferably FOA Certified) and experienced in the tasks they will be doing. Good installers will make the installation easier, faster and cheaper because they will make fewer mistakes. Because of the cost pressure on FTTH service providers, contractors are often chosen by price and then they often subcontract to cheaper, lesser skilled contractors. There have been instances where poorly trained installers, even landscape contractors, have been hired to do installations and have cut other fiber optic cables, punctured water mains flooding neighborhoods or even breaking gas mains causing explosions. Choosing wisely is important.

Focus On FTTH PONs

Most FTTH networks are based on a PON network. The drawing below defines the network: a "feeder" cable extends from the OLT (optical line terminal) in the CO (central office) to a FDH (fiber distribution hub) where the PON (passive optical network) splitter is housed. It then connects to "distribution" cables that go out toward the subscriber location where "drop" cables will be used to connect the final link to the ONT (optical network terminal).

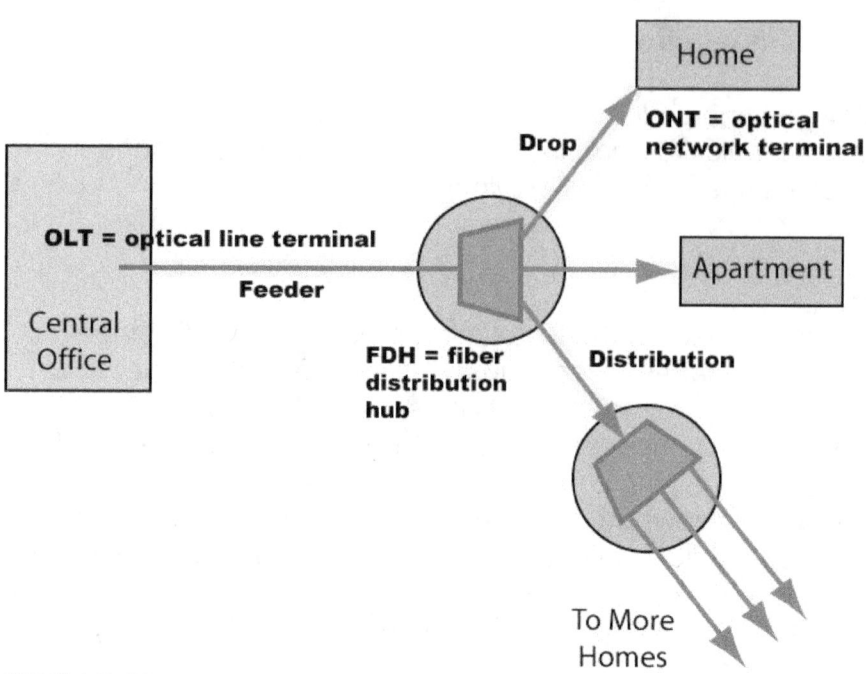

FTTH PON architecture and jargon

The installation of the cable plant to the point where the drop goes to the subscriber is basically standard outside plant construction and outside plant installation which you should be familiar with for FTTH installation. Installation of feeder and distribution cables generally follows standard OSP practice, but the drop cables are unique to FTTH. The cables are often different from normal OSP cables because they have fewer fibers (1 or 2 generally) and are more often factory terminated for plug-and-play at the subscriber interface for single family homes. MDUs (multi-dwelling units) will generally follow fiber to the building conventions with patch panels inside the building.

Preterminated or prefab cables became popular when FTTH service providers realized that they could eliminate the need for experienced splicers at the subscriber installation. Instead, the home installation tech could simply plug in the drop cable, hook up the ONT and connect customer devices. They would only get involved with running cables inside the house if necessary.

Prefab cables can be factory terminated on one end or both. If the cable plant used these drop boxes for prefab cables (below), the drop cables will be terminated on both ends and excess cable will be stored in service loops. Installation of the cable is simply attaching aerial cables, pulling cables in conduit, or using simple trenching techniques to bury the cable in the subscriber's lawn. Keeping buried cables close to sidewalks and driveways minimizes the possibility of them being dug up. And like all underground construction, the installer needs to be aware of any underground utilities in the subscriber's yard, especially sprinkler systems or invisible animal fences which are often poorly documented.

FTTH prefab cabling - Closeup of the six-port drop.

Some special FTTH fiber closures for drop cables require terminating the drop cable to connect it to the box. Patching with connectors in a re-enterable closure has become a popular option to splicing as it allows adding new drops when needed. These closures generally use splice-on connectors, either mechanical or fusion splices, on the bare end of the drop cable. This minimizes the problem of storing excess lengths of cable in service loops. The drop cable can be installed at the subscriber end to the closure then terminated, eliminating most of the excess cable storage.

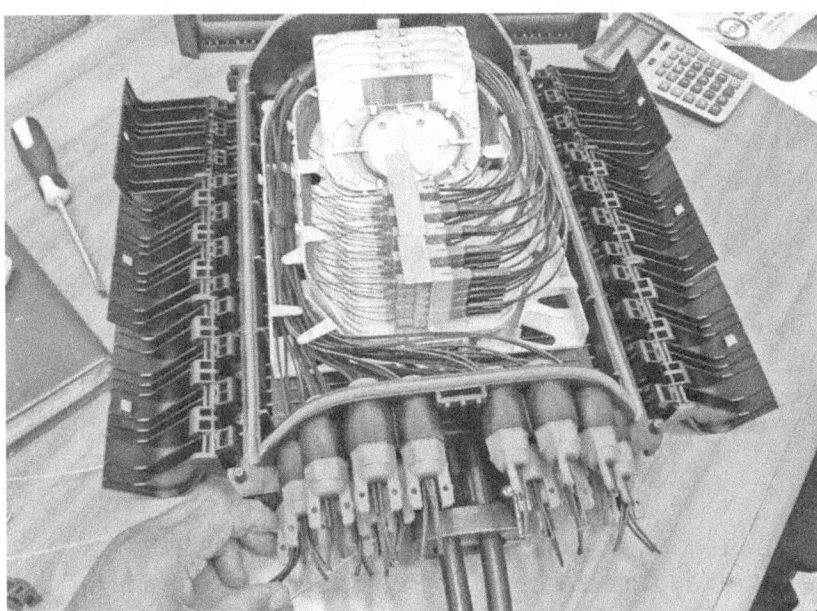
FTTH Drop Cable Closure

This closure has entries for distribution cables, including one coming in and one continuing on to another closure for daisy-chained cables. There are multiple outputs for drop cables which are terminated in connectors. Some closures like this one have provision for splicing on pigtails to terminate the distribution cables while others are designed for direct termination using splice-on connectors using either fusion or mechanical splicing.

If the design calls for termination at the customer premises, these same splice-on connectors are generally used to get reliable terminations quickly. The mechanical spice-on connectors require special tool kits and some practice to get good yield. The fusion splice-on connectors require a fusion splicer but several are available at costs not much greater than the tool kits for the mechanical splice types and require less skill to get good yield.

We're focusing on the fiber part of the installation, but the FTTH home tech will have to know how to connect fiber, set up the ONT and connect all the subscriber's devices. That is a very different skill from fiber optic installation.

FTTH - Customer Premises Installation
Installing a FTTH network is mostly straightforward fiber installation with some additional components like splitters, an ONT and associated hardware at the customer premises that need to be considered. It's when you get closer to the customer premises that things change. Concerns go from laying cable, splicing and installing/provisioning networking equipment to the issues of getting the PON connection to the customer premises and connecting customer devices - while making a positive impression on the customer.

There are issues we have already dealt with in the FTTx section on MDUs involving how to design and install the PON components inside a building with multiple premises. But besides the building installation issues, there are issues when bringing the connection into the customer premises, issues that are similar to those encountered in installing to single family residences. That will be our focus in this section.

Besides the issues of how to physically make the connections, there are important issues dealing with the customer that are probably the single most important part of satisfying the customer. Those issues go beyond figuring out how to make the installation work, it includes dealing with the customer directly, making the installation neat and ensuring that all services work properly.

Dealing with the customer
First impressions count - and are lasting impressions. Personnel doing customer premises installation must 1) dress neatly (preferably in logo shirts or uniforms for the service) and stay clean, 2) have identification (picture ID preferably), 3) be friendly and courteous (remember you are a guest in someone's home), 4) show respect for the customer's premises and 5) minimize the disruptions and modifications that may need to be made. Even the tech's vehicle needs to look professional and identify it as a company vehicle.

After the greeting and first impression, it's necessary to inspect the premises to determine where to install the ONT, the proper way to bring in the cables and how to make connections to the customer's equipment. Remember that practically every installation is unique (although there may be similarities in MDU or subdivision layouts) and every customer will have opinions on where and how to install connections.

Know Local Codes

One must never forget that installing service in the customer premises must follow local building and electrical codes. This includes location of cables and services, avoiding electrical service, sprinklers, invisible pet fences, etc. where installed, etc. All customer premises techs need some training on these codes and know how to follow them in every installation.

Providing A Variety Of Services

Installations may involve installation of phone, Internet and TV services or some combination thereof. Some providers may also offer installation for home networks, security systems, home theater, etc. The work order for the installation will tell the tech what services have been ordered but it's important to check with the customer what services they are expecting before deciding how to proceed with the installation.

Connecting The ONT

The home interface for a FTTH PON network is called an ONT. The ONT may be situated inside the demarcation box for the system mounted on the outside wall of the house and connected to services through the wall. Some ONTs look like cable modems and will require a fiber optic cable from the demarcation box outside into a location inside the house. That type of ONT requires running a cable from the outside into the house to the location of the ONT indoors, similar to connections for other services. Some FTTH service providers use a system called MOCA that allows using the customer's coax cable for CATV or satellite as a network cable for connecting inside the house. Each service provider will have to choose the type of hardware to use on their system, but most will require some installation of cable inside the subscriber's home.

Internet Connection

Connections to the ONT will be determined by the location chosen for the ONT. The connection to the customer's equipment for Internet service will depend on what the customer currently has installed or what they want to have after the FTTH installation. The ONT will provide for a copper cable connection (Cat 5e/6) and/or wireless (WiFi).

The simplest installation could involve connecting the ONT by cable to the customer computer. But most subscribers have several computers or mobile devices, most connected on WiFi. For WiFi, there is a need for a router and/or wireless access point that will be connected to the ONT. Techs should know how to set up WiFi and test various locations for adequate signal strength. Some homes may require building a WiFi mesh network at additional cost. Some homes may also have wired networks connected to a switch that will be connected into the ONT/router.

Phones

Some, perhaps many, premises will no longer have landline phones because the residents have converted to mobile phones and abandoned landlines. But others still have a landline connected into the home phone wiring system. Some users will still have POTS (plain old telephone service using analog connections on a current loop) while others may have been using DSL or VoIP connections. DSL will be replaced by the FTTH connection, of course, and VoIP will be connected into the Internet router provided with the FTTH connection.

Television

Customers ordering TV service may be current users of broadcast TV only, cable or satellite TV. Most users today will be streaming video from the internet, called OTT (over the top.) Users who get their TV signals over the Internet using devices like ROKU, Apple TV, Chromecast, or smart Interconnected TVs. Connections for TV may be made using the customer's current CATV/satellite cable or running coax cable for the TV. OTT users will connect over the Internet, generally using WiFi but may use an Ethernet cable.

Installing Cable And Hardware

Drop Cable

The final cable to the customer premises is called the drop cable. This connection may be aerial, underground in conduit or direct-buried through the customer's yard.

Aerial cable may be self-supporting or lashed to a messenger other cables (CATV coax or telephone) that come into the house. The cable must be properly supported and anchored at the house to withstand typical weather conditions (wind, rain, ice and snow.) In some homes, entering through the attic and going down inside walls may be the least disruptive installation.

Underground conduit to pull the cable to the house may be available in developments that planned for FTTH already or other cable may be pulled out of existing conduit and replaced by the fiber optic drop cable.

If the cable needs to be buried in the customer lawn, the installer or installation crew needs to work with the customer to find an acceptable route and locate services coming to the house first to prevent digging up current services. Buried cable requires more time and (especially) care to minimize the damage to the customer's lawn. Often running alongside a driveway or walk helps minimize damage.

All installations require cable(s) to enter the house at some point. The installer should try to find locations where current services enter the house and use those locations as they are generally more accessible and acceptable to the customer. Where cables enter the house or where the ONT is installed outdoors must be protected and sealed to prevent moisture (or creatures) entering the house.

Drop cables may be pre-terminated and ready to connect to the ONT or require connections on the ONT end and splicing or connecting on the service provider end. The method will generally be chosen by the service provider based on their assessment of the best and fastest method of installation. Installers must be trained in the proper splicing and termination processes to be used in the installation.

ONT

The ONT is the equipment that connects the home to the PON network and provides the electronics for conversion to phone, Internet (Ethernet is the protocol) and TV signals. The first concern is where to locate the ONT. Typical locations will be on the outside of the house, inside a garage or inside a room in the house. Sometimes the location is determined by the service provider, determined by where the incoming cable connection needs to run but the customer may have options, especially as to the location indoors. If the customer already has Internet (CATV cable modem or phone company DSL), the indoor ONT should generally be placed at the same location to take advantage of any current cables.

ONT mounted on an outside wall

Some houses have been built Internet/FTTH ready and have a place indoors already set for the ONT, and, conveniently, have all cables ready to connect. Then the installer only needs to bring the fiber into the house and terminate it, install the ONT and backup power and connect services. Where the ONT is installed indoors, adequate space is required for it and a source of AC power needs to be nearby. If backup power is provided, space for the backup power supply if it is a separate unit is needed. Some ONTs include backup power inside the unit.

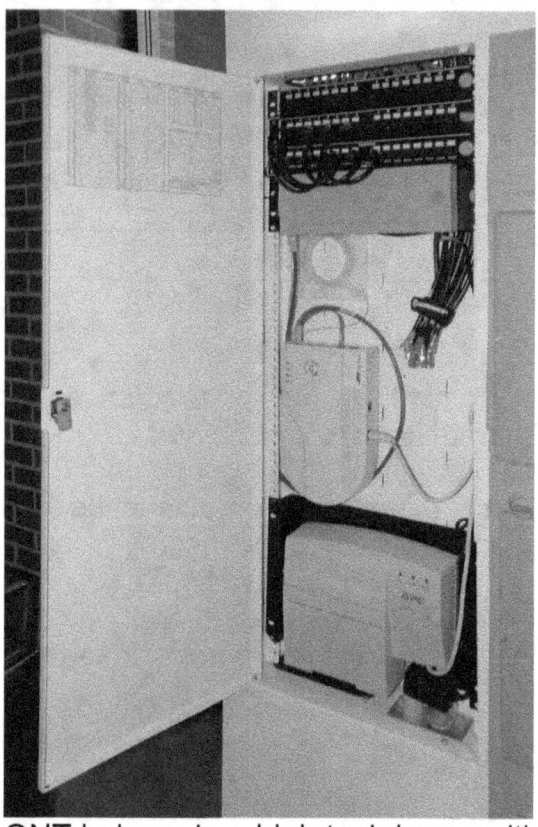

ONT indoors in a high tech home with a wiring panel and room for the ONT and backup power

If the ONT is located outdoors, the connections into the home will be the connections to the customer's equipment. If the ONT is located indoors, the installer will have to bring a fiber cable inside the house to the location of the ONT. Either way, connections will need to penetrate the outside wall of the house. Great care must be taken to avoid other services when drilling holes and the penetration must be properly sealed.

Cabling To User Equipment

The ONT provides cable connection to phones (POTS) and the Internet (Ethernet) over modular jacks for unshielded twisted pair copper cables and to the TV with coax cable. Ideally, the location of the ONT will be where earlier service provider equipment was located and current customer cables are available at that point. If not, it may be necessary to install new copper cables

Cabling Inside The House

The first thing to remember about cabling inside the house is to keep the customer informed and involved in the process. People are very sensitive when installers start poking holes in their walls and snaking cables. Always tell the customer what you are doing and why. Always get their input on where to locate equipment and cables and where to penetrate walls, floors or ceilings.

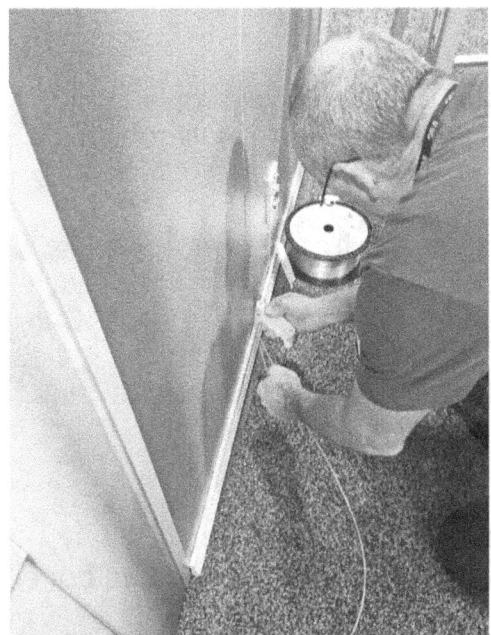 

Manufacturers have solutions to make cable installs inside the customer premises simpler and easier - and almost invisible

Always be neat and clean up after your work.

Training The Customer

After the installation of hardware is complete and your services are installed, show the customer what you have done and train them on how to use the services you have installed. This should include a demonstration of the phone, Internet service and TV. Show them the instruction materials and how to contact customer service. Instruct them on the Internet connection, speed testing, email services, and the service provider gateway. For the TV, instruct them on the use of the remote, gateway services, channel selection, etc. If you provide a multi-gadget remote, program it for their TV, DVR/VCR, DVD player for them if you can.

Above all, make sure they know how to contact customer service so be certain that information is easy to find in the materials you leave them.

Finish your cleanup and wish them happy use of the system before you leave.

Testing The Installation

The FTTH PON cable plant is a complicated testing problem with multiple connections, wavelengths, and bi-directional splitters which we cover in the next chapter. In some cases, the installation is simply tested with a fiber optic power meter which checks to see that the optical power level at the users' ONTs is in the proper range. If it is, the cable plant is likely to be within spec and the system should operate properly. If not, troubleshooting will require a technician with knowledge of PON systems and how to troubleshoot them.

FTTx Safety Issues

FTTx safety issues include all the usual fiber installation issues, for example construction and installation of the cable plant, working with bare fibers, solvents and adhesives. But FTTx networks have several other potential problems.

BPON links carrying AM CATV signals may have high power from EDFAs, especially before the splitters. And links may have multiple equipment transmitting simultaneously. Either case can cause high optical power that can be dangerous to worker's eyes. Care should be taken to not expose eyes to light from the fibers and to always use microscopes with infrared filters, just in case. Since systems may have multiple systems transmitting on the same fiber, it is harder to ensure that all systems are turned off for inspection or testing, also.

And, since 32 or more users may be sharing the CO based network equipment, turning off systems for troubleshooting is not desirable, so testing may have to be done with equipment in service. Exercise care.

All FTTH projects must follow normal safety practices for construction and installation as defined by organizations like OSHA (US) or the equivalent agency for workplace safety in your area. More on fiber optic safety can be found in the FOA Guide. Look for sections on Safety as well as the OSP Construction section.

Chapter 6. Testing FTTH Networks

Objectives: From this chapter you should learn:
Special needs of testing short PON networks
How some fiber optic tests require different procedures

Introduction To FTTH Testing

Network architectures (PONs or passive optical networks) have been developed that allow sharing expensive components for FTTH. A passive splitter that takes one input and splits it to broadcast to 32 or more users cuts the cost of the links substantially by sharing, for example, one expensive laser with 32 homes and only requiring an inexpensive laser at each home. However, this architecture changes the methodology of testing the complete installed cable plant and links for proper operation. Of course, individual links are tested as usual, it is the PON coupler that creates the difference.

A PON cable plant is a complicated testing problem with multiple connections, wavelengths and bi-directional splitters. In some cases, the finished installation can be tested with a fiber optic power meter which checks to see that the optical power level at the users' ONTs is in the proper range. If it is, the cable plant is likely to be within spec and the system should operate properly. If not, troubleshooting will require a technician with knowledge of PON systems and how to troubleshoot them.

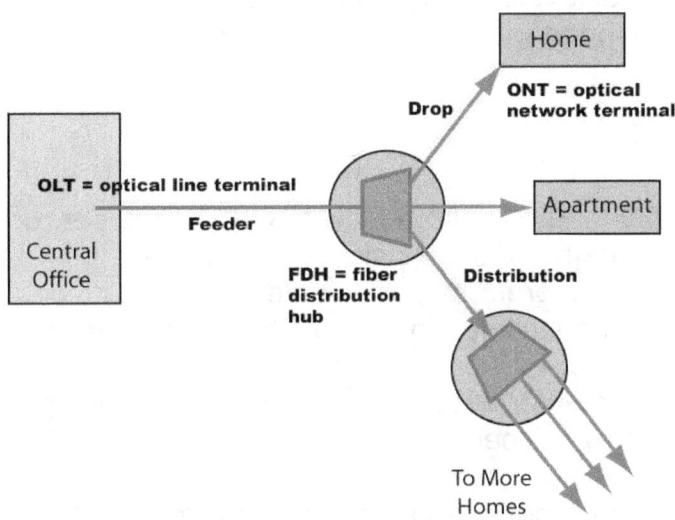

FTTH PON

Each subscriber needs to be connected to the local central office with a single singlemode fiber, through a local PON splitter (or maybe two if the PON splitters are cascaded.) Every home will have a singlemode fiber link pulled or strung aerially to the phone company cables running down the street and a network interface device containing fiber optic transmitters and receivers will be installed on the outside of the house. The incoming cable needs to be terminated at the house, tested, connected to the interface and the service tested. See the FTTH Architectures chapter for more information on typical FTTH installations.

FTTx Testing Issues

Testing FTTH network is similar to other OSP testing but the splitter and WDM add complexity. FTTP PON networks can be more complicated than simple OSP links, with WDM couplers, PON splitters, etc. in a single link, so complete testing can include some components and installation issues not familiar to the usual OSP tech. PON couplers add high loss, WDM couplers have different performance at different wavelengths and connector reflectance, not a problem in most systems, can be a problem in short links typical in FTTx. Many FTTx systems use APC (angled PC) connectors to reduce reflectance so test cables for both OLTS and OTDR need to have matching connectors.

However, once installed, the active users on a live network means testing cannot disrupt service. Thus, testing may be as simple as checking power at the ONT on the subscriber's house with a calibrated fiber optic power meter or just seeing if the ONT has a "green" connection light! The ONT at the home usually has some intelligence that can be accessed from a remote location, allowing a service tech to initiate a loopback test to verify connections at any user. If only one user has a problem, a service tech is then sent there, while if all users are down, the tech is sent to the central office.

As with most fiber optic links, troubleshooting requires knowing the architecture of the system, expected link losses and optical signal levels and typical problems that may be encountered. As always, we emphasize the importance of having documentation on the system before testing and troubleshooting.

Link Testing

A link is a single run of fiber, e.g.: from CO to FDH or from FDH to ONT. The fiber run may have connectors or not, depending on whether the links are spliced or use connectors for terminations. Quite a few now use preterminated cables to speed installation. The loss of the PON splitter must be included in the loss budget for the link.

See the FTTH Architectures chapter for more information on PON splitter losses. If you need to test just the splitter itself, directions are below.

You must measure loss with OLTS at all wavelengths and bidirectionally to check all operational modes - similar to how the transmission equipment will use the fiber.

The installer may need to characterize each fiber with an OTDR, verifying fiber attenuation, termination losses and reflectance and splice quality. The OTDR will also show any bending losses caused during installation. OTDR traces should be filed for future reference.

Optionally, the installer may test splitters at the FDH or the WDMs at the CO. If these are pretested, as they should have been, this may not be necessary or advisable, especially since it is time-consuming and costly. WDMs also require specialized test equipment.

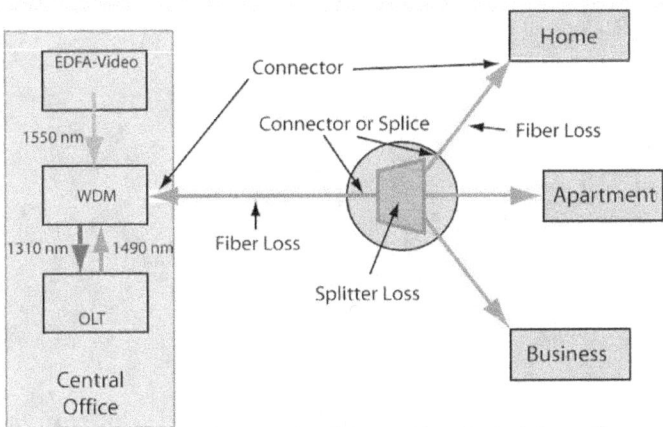

FTTH PON loss testing requirements

After the link is installed, it needs testing from end to end. The end-to-end loss includes the connectors on each end, the loss of the fiber in each link, the connectors or splices on the splitter and the loss of the splitter itself. Since the fibers are being used bi-directionally and connector or splice loss may be different in each direction if the fiber core diameter (mode field diameter for SM fiber) is different, testing in both directions is important too. Special FTTx PON OLTS are available that test the proper wavelengths in each direction, simplifying testing logistics.

Since PON links are generally short (<20 km) chromatic dispersion (CD) and polarization mode dispersion (PMD) are not a concern. CD and PMD are generally only issues on very long high speed links.

Let's consider the most complex version of PON testing, BPON. It's similar to any OSP testing but the splitter adds much more loss and WDM adds complexity since there are three wavelengths in use. Tests include each splitter, each link and end-to-end loss. Loss and reflectance are especially important if systems are using an AM video transmission system at 1550 nm, as it has a maximum tolerable loss and reflectance before signal quality is noticeably affected.

Tests may need to be done at all three wavelengths of operation: 1310 nm for upstream digital data, 1490 for downstream digital data and 1550 nm for AM video downstream (BPON) but testing at either 1490 or 1550 nm should suffice for both those wavelengths.

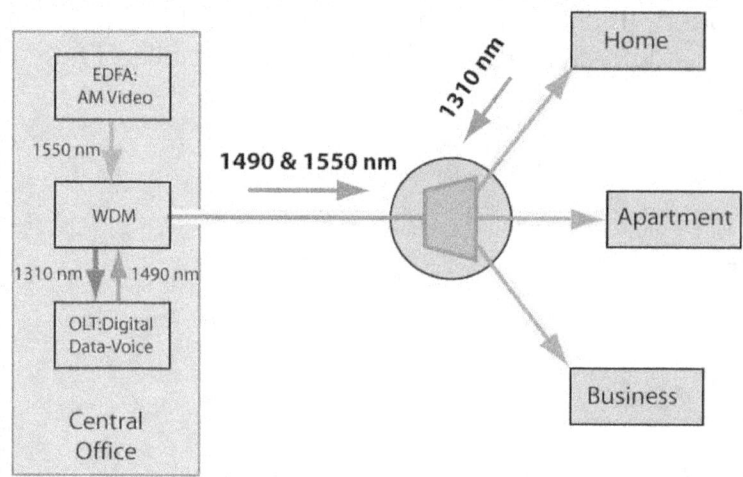

FTTH PON wavelengths used in BPON

Insertion loss of the cable plant including the loss of the coupler is tested using an optical loss test set (special test sets for FTTH PONs are available that cover all 3 wavelengths of interest.) OTDRs can be used if length is adequately long, to determine connection reflectance, fiber attenuation and troubleshoot problems. Many systems will take OTDR traces and store for troubleshooting. The splitters can confuse the OTDR so one generally reverses OTDR test, taking traces from the subscriber upstream.

OTDR Testing With PONs
Using an OTDR to test every fiber in an OSP link is traditional, as the OTDR provides a snapshot of the losses in the fiber, locates loss events (connectors, splices and bending losses from improper installation), aids installation troubleshooting and provides a trace which can be stored for later troubleshooting and restoration.

OTDRs work well with long haul fiber optic cable plants, documenting the loss of the fibers and splices connecting cables. However, OTDRs are less useful in short links or in complicated optical networks like PONs.

In FTTH PON networks however, the lengths of fiber in the cable plant are often very short and the PON splitter produces some unusual traces on OTDRs, with the traces looking totally different when tested from each direction. Interpreting these traces can be difficult. Because of these difficulties of testing with an OTDR, the FTTH PON cable plant is often just tested with an optical loss test set and not an OTDR. However, we will explain the issues with OTDR testing on a PON network to help users understand the methodology needed for OTDR testing.

Here are two traces from an actual system taken in two directions.

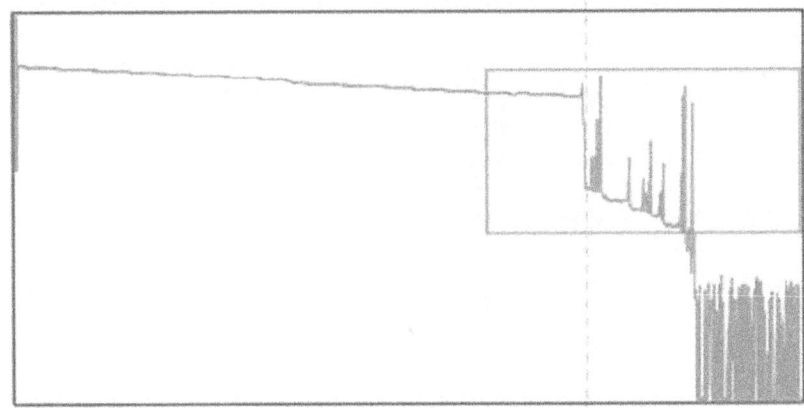
Trace taken downstream from the CO (OLT) to the subscriber (ONT)

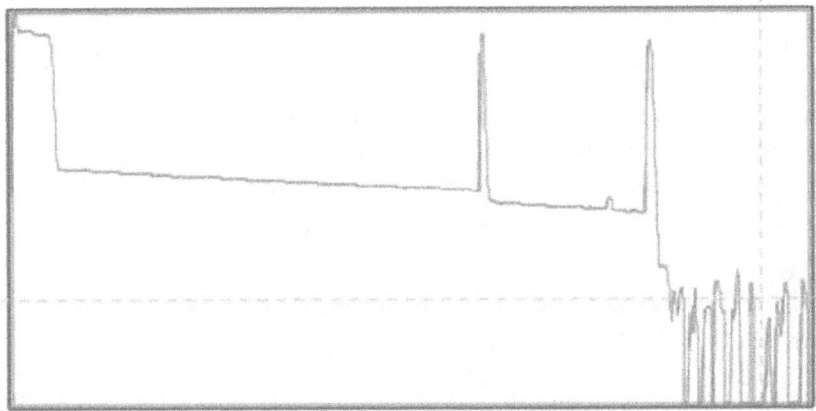
This trace is taken upstream from the subscriber (ONT) toward the CO (OLT)

In both traces, you can see the large loss of the PON coupler, best seen in the upstream trace at the bottom, on the left side of the trace. On the downstream trace, it is the large loss preceding the multiple peaks of the subscriber fibers, marked with the gray rectangular marker. Below we will show a simpler coupler and explain what you are seeing here.

OTDR Testing From The CO/Head End (OLT)
  PON systems create problems for OTDRs. Shooting from the input of a PON splitter at the CO (OLT), the OTDR trace will show the fiber to the splitter as a normal trace, then it shows the loss of the splitter, then the combined traces of all the connected fibers after the splitter. The OTDR sees and adds together the backscatter traces from all the fibers after the splitter, making it impossible to determine the trace of any of the fibers individually. Since it is impossible to see detail on individual fibers, any event associated with that fiber (connector, splice or bending loss) cannot be easily assigned to any individual fiber.

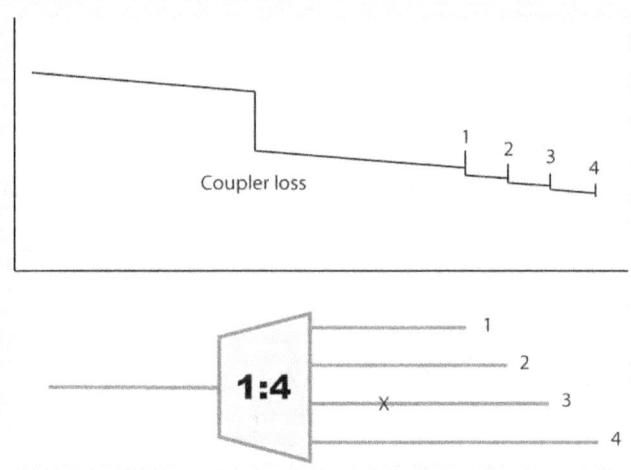

Testing PON with OTDR downstream

Consider the "X" shown in the splitter diagram above. If it was a loss or reflective event, it would show on the OTDR trace, but the operator would not know if if were in fiber 1,2,3 or 4. The only unambiguous part of the OTDR trace shown is the end of fiber 4, the longest fiber, beyond the length of the next longest fiber, #3.

It should be noted that FTTH links, because of their short lengths and the use of some high power transmitters, usually have APC connectors or fibers prepared to have minimal reflectance. That can make analyzing downstream OTDR traces very difficult when no reflective end is available to mark the fiber end and there are 32 or more fibers in the system.

Here is an illustration of how a real trace can become very complex to analyze. This is an enlargement of the coupler to subscriber section of the downstream trace above which is outlined in red on the trace.

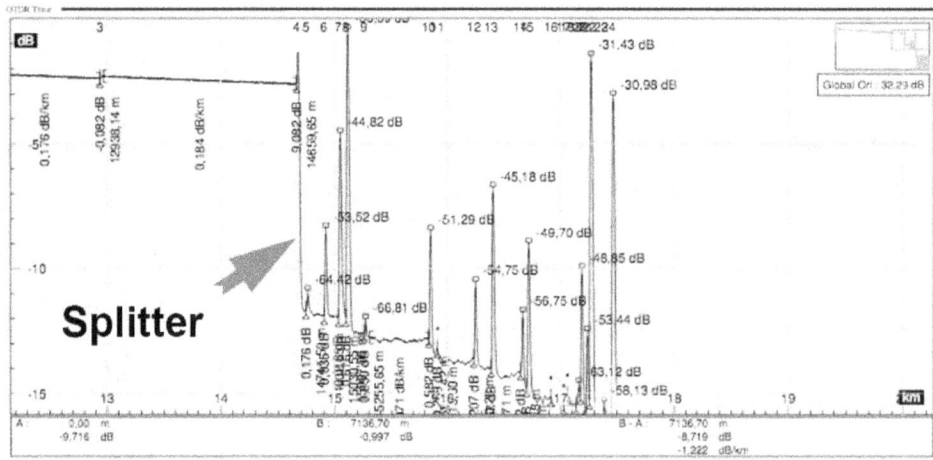

FTTH-OTDR-downstream trace showing all the combined fibers after the splitter (coupler)

As a result of the complexity of downstream traces, OTDRs are generally used on PONs from the subscriber (ONT) end toward the CO (OLT) to characterize the fiber path. However, the OTDR may also be used from the CO end, but, as you can see from the diagram above, it only allows the operator to characterize the length of each fiber link, providing actual fiber length to add to network diagrams for future troubleshooting.

Special PON OTDRs will test at 1310, 1490 and 1550 nm. Some also test "out of band" at 1650 nm, which is more sensitive to bending losses and allows in-service testing with a filter to remove signal wavelengths. Since PONs are short, the OTDR needs very high resolution, usually obtained by having the shortest test pulse that will give adequate range.

Testing PONs in the downstream direction is helped with launch and receive cables. The launch cable allows testing the initial connector on the link as well as allowing the initial overload of the OTDR to settle down as with any OTDR test. But on the receive end, if a cable of known length is used, say 100 m or 500 m, one can look back exactly that distance from the reflective end to see the loss of the end connector.

OTDR Testing From The Subscriber End

Testing from the subscriber end is easier. The fiber path will show events on just one fiber, like the "X" shown on fiber 3, and a high loss for the coupler. Here a 1:4 coupler will have 6 dB of splitting loss plus perhaps 1dB excess loss for a total of 7 dB loss. Using launch and receive cables allow testing connectors on both ends and measuring end to end loss.

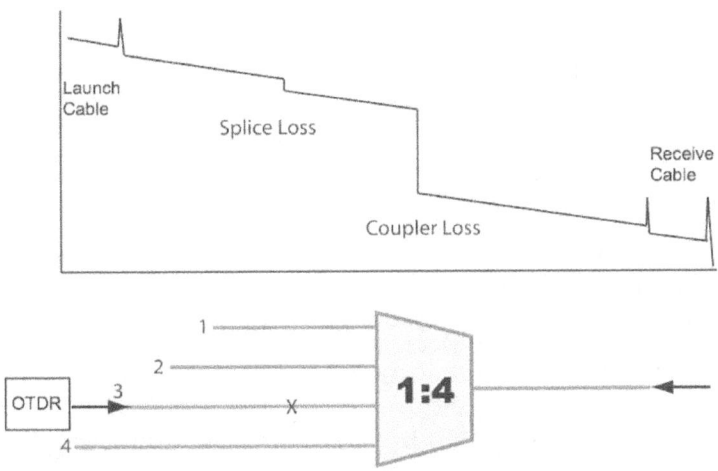

Testing PON with OTDR upstream

Here is a detailed trace from the upstream example above, showing how much simpler the trace is when the other subscriber links are not shown.

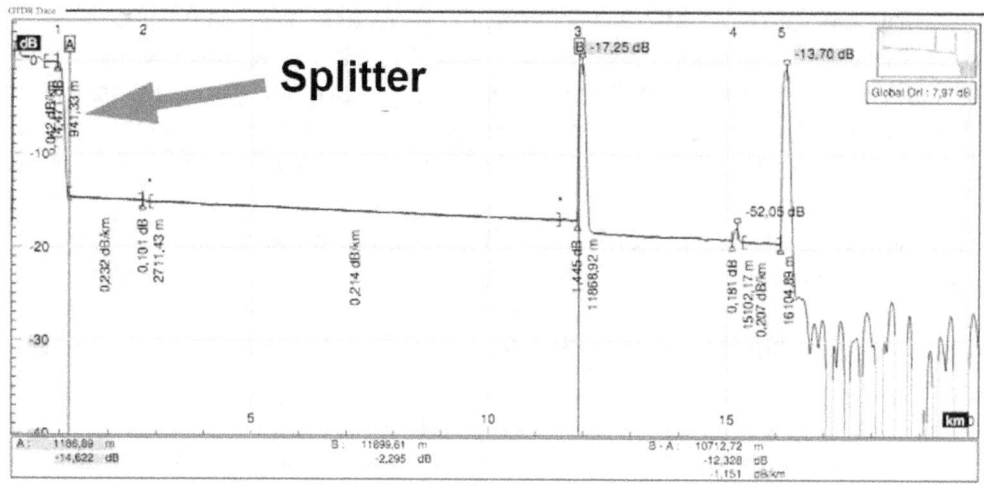

FTTH-OTDR-upstream

Testing Splitters

Splitters are tested following the same directions for a double-ended loss test. Attach a launch reference cable to the test source of the proper wavelength (some splitters are wavelength dependent), calibrate the output of the launch cable with the meter to set the 0dB reference, attach to the source launch to the splitter, attach a receive launch cable to the output and the meter and measure loss.

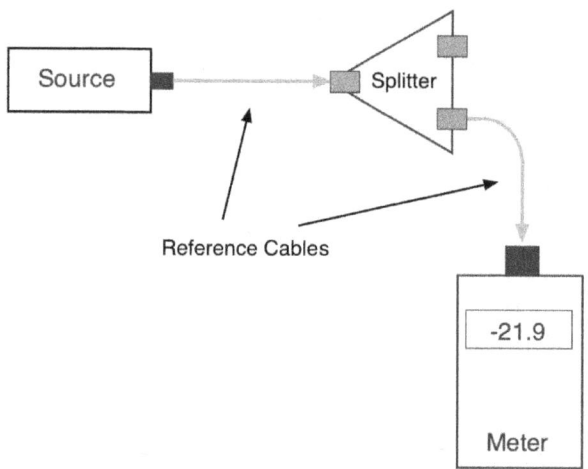

Testing a PON splitter using a light source and power meter.

What you are measuring is the loss of the splitter due to the split ratio, excess loss from the manufacturing process used to make the splitter and the input and output connectors. So the loss you measure is the loss you can expect when you plug the splitter into a cable plant. You can test the splitter as shown in the diagram above to also determine the loss for each output to see how consistent the outputs are. Likewise you can test in the opposite direction to see how it works in combining signals.

Network Equipment
Network equipment will be tested as the system is turned on or for troubleshooting. Will the network equipment transmit and receive properly? If the cable plant is installed correctly and tests within specifications for loss and reflectance, it should. Most FTTx equipment has extensive self-testing capability and that may prove sufficient for most testing. PON couplers may have a second port on the upstream side just for testing or unused downstream connectors may be useful for testing, especially with OTDRs. And the subscriber side of a splitter usually has several unused ports for future expansion that can be used for testing.

The network equipment should be tested for optical power. The transmitter output should be within specifications, as should the receiver input, when tested with a calibrated optical power meter set at the proper wavelength(s). If testing is done while all three systems are operating at their respective wavelengths, a power meter with wavelength selective input is required. Power at the receiver is critical. Too low and the signal-to-noise ratio will be too low; too high and the receiver will saturate. Both conditions will cause transmission errors. High power is not uncommon, so attenuators may be used in these links to reduce power to acceptable levels.

Data transfer testing with a protocol analyzer is the final test. It will be done using specific protocol testers for the data formats being transmitted. Personnel doing these tests are probably not the same that test the cable plant as each have specific training and test equipment needs.

ONTs are generally capable of loopback testing under remote control which does not require any test equipment. This may mean more sophisticated testing is unnecessary for troubleshooting.

FTTx Safety Issues When Testing
FTTx safety issues include all the usual fiber installation issues, for example construction and installation of the cable plant, working with bare fibers, solvents and adhesives. But FTTx networks have several other potential problems.

BPON links carrying AM CATV signals may have high power from EDFAs, especially before the splitters. And links may have multiple equipment transmitting simultaneously. Either case can cause high optical power that can be dangerous to worker's eyes. Care should be taken to not expose eyes to light from the fibers and to always use microscopes with infrared filters, just in case. Since systems may have multiple systems transmitting on the same fiber, it is harder to ensure that all systems are turned off for inspection or testing, also.

And, since 32 or more users may be sharing the CO based network equipment, turning off systems for troubleshooting is not desirable, so testing may have to be done with equipment in service. Exercise care.

All FTTH projects must follow normal safety practices for construction and installation as defined by organizations like OSHA (US) or the equivalent agency for workplace safety in your area. More on fiber optic safety can be found in the FOA Guide. Look for sections on Safety as well as the OSP Construction section.

Chapter 7. Case Studies - Do It Yourself FTTH

Objectives: From this chapter you should learn:
Case studies: How 2 groups wanting FTTH built their own networks
What they learned from their experiences doing it themselves

Introduction

Do you think that creating a FTTH network is only possible if you have the resources of a Verizon or Google? That you need a contractor with lots of experience in designing and installing the fiber network? Or an IT department who can install and operate the equipment? Well, think again.

With most people agreeing that broadband Internet is an essential utility, the problem becomes how to get it in areas that incumbent service providers don't have any interest in serving because of the cost? More and more groups are deciding to do it themselves. DIY FTTH is completely feasible, and hundreds of organizations are doing it already. These DIY FTTH projects are being done by cities and towns, utility coops, especially electrical coops who need fiber for their grid management anyway, as well as private groups like homeowners' associations, real estate developers and even private companies with venture capital funding.

Perhaps the best known is the Electricity Power Board of Chattanooga, TN, EPB has shown that gigabit broadband can transform a sleepy city into a booming manufacturing (VW factory) and tech city. But you don't need to be as big as EPB to bring broadband to your area as these two examples show.

Southern FiberWorx

About 2014, FOA was contacted by Greg Turton of Cordele, GA. who was curious about what was involved in creating a FTTH network. Greg is a real estate developer who also owns several local hotels. Where he lives and builds homes is way outside of a service area that anybody wants to build good broadband, forget FTTH. Cordele itself has a population of only about 15,000 and is one of those small cities along the Interstate highway that are everywhere in the US.

We answered Greg's questions and led him to some of the FOA Guide web pages and YouTube videos about FTTH to get him started. More conversations discussed how to get connections as an ISP, types of components and suppliers, etc. Fortunately, the local electrical utility has lots of fiber but they were restricted from building their own FTTH network because Georgia was one of 19 states where lobbyists for the incumbent providers got laws passed restricting their ability to operate a FTTH system themselves. But they were more than willing to lease dark fiber to Greg at really good rates. And there were good choices on getting an Internet connection. As he got more serious about the project, we introduced him to two FOA Master Instructors, Eric Pearson and Dominick Tambone, in Atlanta, just two hours away.

Greg hired Eric to come to Cordele and train him and several more of his people. Eric taught them how to work with cable, prepare the cable and splice it, dress cables in splice closures, pedestals, manholes, etc. Eric, Dominick and the FOA had many conversations with Greg about his project and the potential suppliers to it. As construction began, Dominick came down to Cordele to help with the early installations.

Now that you know where we're going with this story, let's talk about Greg. He is not your usual fiber optic project manager. First he is a second generation developer, following in his father's footsteps. He is accustomed to getting into the depths of a project, understanding the risks and making investments. He's also a "tinkerer" - he likes projects and challenges. He's a private pilot who has invented and manufactures two types of air conditioners for small airplanes which he built in his shop and tested in his own airplane. He makes electrical hardware he invented for his own hotel to make using tech devices more convenient for his guests. He also has a great crew of people working for him and lots of local connections.

To Greg, FTTH was another challenge, and after he became trained himself, he knew he and his crew could build it themselves. And they did!

Greg and his crew had a lot to learn and accomplish before the project even started, as we mentioned above. They also had to figure out how to document the system as they designed and built it, something they accomplished using Google Earth.

The first think Greg lined up was an Internet connection for his project. Cordelle was not far from a major highway which had multiple fiber optic cable running alongside it. He was able to arrange a connection but had to also arrange for a fiber optic link to his office where they would have their head end, which involved getting permissions and permits for the link.

As they started buying equipment for splicing and testing, the needed a splice trailer. No problem, they built one. Based on their training, they knew they needed a splicing trailer for work in the changeable South Georgia weather. A small travel trailer provided the base for their splicing trailer, and they furnished the inside with a work counter and racks and cabinets for storage. Air conditioning was furnished by a local company that wants a fiber connection themselves.

Rather than investing in heavy equipment, they rented what they needed from local companies. They also converted a large utility trailer to carry cables and conduit.

Most cables were installed by trenching. To avoid digging up paved roads or driveways, they used directional boring. It also worked well for several bridges crossing small streams.

In an area that has lots of underground utilities, they had to be very careful. They started by calling utilities before they started and manually digging holes before using the backhoe. Finding after a few near misses that such data wasn't sufficient, they bought an underground locator and learned how to use that to double-check before digging.

Splice closures were put into hand holes or pedestals like this one. The cable was laid along two lane roads along with other utilities. Greg is shown above with one of the equipment pedestals needed to get connections into a remote area.

Greg at one of the system pedestals

Here's Greg and David Herlovich, his assistant, with the equipment for their head end. They chose ADTRAN equipment for their system because of their reputation, knowledge, and support, plus they are nearby in Huntsville, AL so Greg can fly his plane over to visit the factory when necessary.

The proof is in the performance, of course. This is a computer at Greg's house connected to his network doing a speed test.

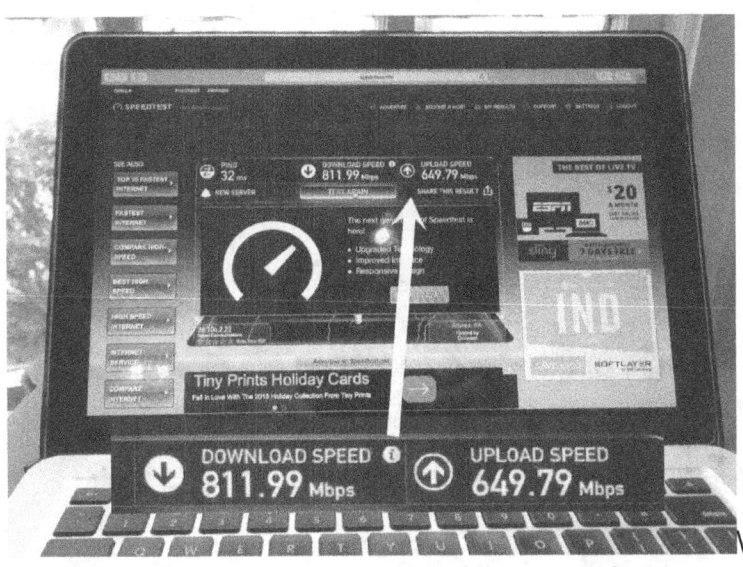

And there you see the results of the speedtest on a "do it yourself" gigabit FTTH network.

When we visited Greg, Southern Fiber Worx had just started installation. They had already connected 30 homes and had over 100 scheduled for installation. His original goal was to pass about 800 houses in his development and sign up 30-40%. Southern FiberWorx works like Google Fiber; get your neighbors together and sign up and they build that neighborhood next. Now Southern FiberWorx is a booming FTTH provider.

While we were in Cordele, we talked to some local businesses and discovered that what Greg had been telling us was true - the local enthusiasm for what he was doing was amazing and people want to get connected ASAP. Because Greg knows practically everybody in town and has talked to many as he got permits and help building the system so far, they know what he's doing and want him to expand beyond his development to cover the entire town. As the word spread, the county expressed similar interest in his expanding the Southern FiberWorx footprint to cover the county. Then the next county approached him with the same idea.

While we were in Cordele, Greg asked us to visit his bankers who wanted to know more about building a FTTH network - you know, from the investment point of view. To date, no kidding, Greg has funded Southern FiberWorx out of his pocket! Yes, it does not cost that much to get something like this started. But if he expands to the city of Cordele and the two local counties, he may need to get financial backers. We pointed out to the bankers that FTTH provides high income with little overhead making good cash flow. In addition, recently two CATV systems had been sold for $5-6000 per subscriber - a whole lot more than it cost to connect each of Southern FiberWorx subscribers on gigabit FTTH. If that's not a good return on investment, what is?

Connect Anza

In late 2014, Kevin Short, General Manager of the Anza Electrical Cooperative called FOA to ask questions about building a fiber optic network. FOA visited Kevin and subsequently met with the Board of Directors of the Coop to discuss ideas about building a fiber network over their electrical network. Their electricity supplier was pushing them to build fiber for grid management (Smart Grid) and it seemed reasonable to assume that once the backbone was built, expanding to provide Internet to their customers was possible.

There was certainly a desire for better Internet because they did not have any. It's easy to understand why. Anza is really, really, rural - located in the Southern California high desert at the southern end of Mount San Jacinto State Park. We dubbed the AEC project "FTTR" for "fiber to the ranch" because of the typical customers in the service area!

Anza gets their electricity from an electric cooperative. With the help of Franklin D. Roosevelt, who established the Rural Electrification Administration in 1936, friends and families banded together to create a new kind of electric utility, where the voice of every person made a difference. Electric cooperatives brought electric power to the countryside when no one else would. Electric cooperatives are owned by their members and focus on their member needs and local priorities.

Anza Electric Cooperative, Inc. (AEC), energized in 1955, is a member of Touchstone Energy® - the national brand of electric cooperatives - providing power to the communities of Anza, Garner Valley, Pinyon Pines and parts of Aguanga. AEC provides power to 3900 homes, schools, and businesses in a 700 square mile area.

AEC's service area is nearly 700 square miles of high desert with an elevation at roughly 4,000 feet where winter weather can sometimes be a challenge. Anza is located at an almost equal distance from Palm Desert, Hemet and Temecula in Riverside County in Southern California.

While Anza is quite rural, it was only an hour's travel from FOA HQ, so FOA President Jim Hayes volunteered to do a series of half-day training session for AEC personnel on fiber optics and installation practices to familiarize them with what they would be doing in the future to learn from the project so we could share it with our readers.

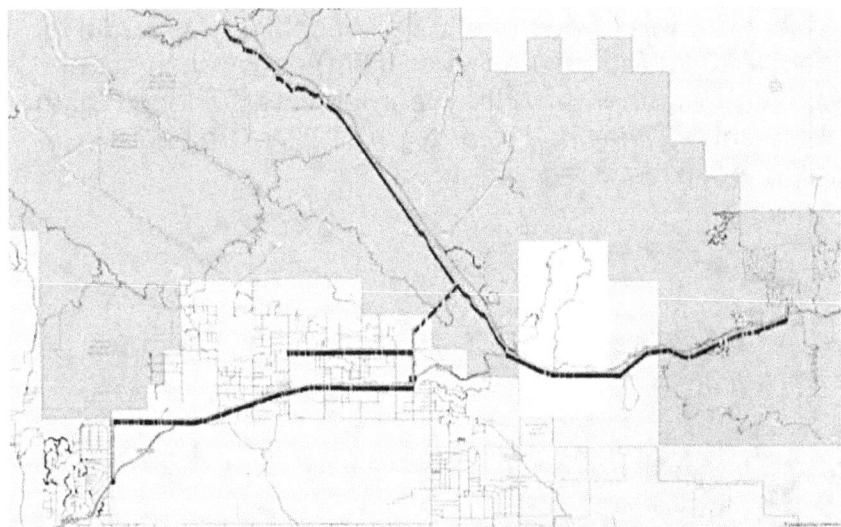
AEC's fiber backbone run along their electrical lines.

AEC applied for and got a grant from the California Public Utility Commission for $2.6 million, about $700 per household, to help pay for the project. One thing is important to understand about rural projects - they cost a lot more than urban or suburban FTTH networks, and the CPUC grant would cover only about half the total cost.

Since AEC is a coop, a bylaw change was voted on by Anza Electric Cooperative members with an overwhelming 91.3% of members approving the bylaw change to include fiber optic, high speed internet service to our members along with our traditional electric service.

Getting Internet service to a very rural small town can be difficult, but Anza had a state highway connecting the California coast to the Palm Springs area. Along that highway was a major telecom link with Internet service running next to their offices, so an Internet connection was simpler than expected.

To do the design of the network, AEC used a unique solution. They enlisted Jeffery Willis, a local resident who was a student at the University of Colorado, Boulder's Interdisciplinary Telecommunications Program, to do the design as expansion of a pre-existing Master's Degree Capstone project that AEC assisted him with. This was a brilliant idea due the success of this college feasibility study. Before getting started, Jeff performed a plethora of research and development on all aspects of the design, including a survey of design software which he shared with FOA for our readers. FOA helped him with some fiber issues, but he had relatively little trouble doing the design for the project. "

Much of the backbone ran along roads in the mountains that connect the various population areas of the AEC service area. Some areas had restrictions on where cables could be run because the roads were designated "scenic routes."

One thing to consider in a project like this - you need LOTS of cables! Anza had to store over a dozen spools of cable – 20 kft (about 4 miles or 6 km) each, each spool weighing about 1600 pounds (726 kg)! Another thing they learned was the length of the backbone was not the length of the cable. They had to order 10-12% extra cable to accommodate service loops, drops for splice closures, etc.

More than 50 miles (80 km) of cable stored in a AEC building.

They also needed more equipment, including this bucket truck made into a rolling advertisement for "Connect Anza."

Anza bucket truck advertises Connect Anza.

Installation of the cable plant for ConnectAnza was somewhat out of the ordinary. The area covered by the coop included some areas without telephone service and poles and/or messenger wires needed to be installed in many areas. One section had to be bridged with wireless because the rod was a "scenic highway" and poles and aerial cable were not allowed!

Splicing was done on the ground, often in the desert.

This closure has provision for a PON splitter and several drop cables.

The head end for ConnectAnza is installed at their main office in the town of Anza. This rack includes the router for their Internet service and the ADTRAN OLT equipment. In the bottom of the rack is the backup batteries, an important part of the equipment for an ISP.

ConnectAnza head end in the Anza Electrical Coop office

ConnectAnza is now a fully functional ISP, one of the most rural systems we know. They are proof that rural FTTH can be built, and it is certainly welcomed by their subscribers. From no service, coop subscribers can get 100/100 Mb/s service (100 Mb/s downstream and upstream) for $49/month, 300/300 for $79/month. Low income residents have a basic service of 20/20 Mb/s for $20/month.

Chapter 8. FTTH Project Management

Objectives: From this chapter you should learn:
What's involved in a FTTH project
Issues in managing a FTTH project
Special considerations of do-it-yourself FTTH projects
Workforce development techniques

Introduction

Most information about fiber optics, including the information in the FOA textbooks and the FOA online Guide, is written for the technician who designs, installs or tests the network. But many times, if not most of the time, the success of a fiber optic project depends on those overseeing the project. This includes the manager of the organization for whom the network is being built, the planners behind the project, financial managers and particularly the people who supervise and evaluate the installation itself. After the project is done, there must be managers and supervisors who ensure the project runs smoothly, delivering the communications that keep users satisfied.

In this section on fiber optic projects, FOA ties together topics covered in many pages in the online FOA Guide and in chapters in some of our current textbooks, to provide a reference for those who manage the personnel we mention above. Here the focus is on the project from conception to completion; managing the people who design, install, and operate it. We provide guidelines for all phases of the project, including enough technical details that managers can understand what technicians are doing and reporting about the project. We recommend linking to the FOA Guide "Jargon" page, a page about the "language of fiber optics," that helps everyone speak the same language when discussing fiber optic projects.

Don't expect this section to provide all the answers; we don't even know all the questions! Every fiber optic project is different and unique. The communications needs, the geography of the cable plant, local laws, codes and regulations, and even the available technology, which is ever changing, will all be unique to your project. Our hope is that we provide sufficient background that you can understand your own project well enough to manage it successfully.

What Is Involved In A Fiber Optic Project?

A fiber optic project begins with a need for communications and ends with an installed fiber optic cable plant and an operating network that fills that communications need. Between those two points are a number of stages:

- Concept
- Selling the project to decision makers
- Getting financing
- Designing the project
- Installing the project

- Accepting the built network
- Operating and maintaining the network

Each of these stages breaks down into many smaller projects with one thing in common - they require a thorough understanding of the project and careful management to ensure the end result is what is expected. This page is aimed to provide resources for the manager of the project and the people building the project to allow them to have a mutual understanding of what will be happening as the project moves from concept to completion.

Focus On The Cable Plant

The FOA's expertise in in fiber optics and we generally focus on the fiber optic cable plant. What is a "fiber optic cable plant"? It's a term we use all the time in fiber optics to cover the installed fiber optics that can transmit your communications signals. It's permanently installed between the two points which you require communications between. It's what you connect your communications electronics to with patchcords on each end.

The cable plant includes all the fiber optic cable between those two points. That cable may be buried underground or installed aerially on utility poles. It may even have segments that run under water - streams, rivers, lakes or oceans. Cables come in a maximum length of about 5-6 km on a spool from the factory, so longer lengths will require splicing cables together. Splicing is also required where points along the route require connections (drops) as well as from end to end.

At the ends, the cable plant will be terminated in connectors to allow making connections that can be changed as needed. Hardware is required for every splice and termination to protect the cable plant, splices, terminations and connected equipment, and these may require underground storage in manholes or above ground storage in pedestals, huts or buildings.

Designing and building a cable plant means carefully and completely defining the entire route of the cable plant, where every splice, drop, termination and piece of hardware is to be placed and what components will be used for every bit of the cable plant. It's a big job to design, but it must be done correctly to allow installation to be done according to the needs of the users.

Once finished, the cable plant must be fully documented so it can be operated, maintained, and repaired if restoration becomes necessary.

Fiber Optic Project Management (For Managers)

The FOA, as part of the fiber optic industry and especially in our role as educators, most of our focus has been training installers of fiber optic cable plants and networks in fiber optics. But what about the people for whom they work or build the networks? FOA is concerned with their education too. What do network managers, project managers, supervisors, network owners, IT personnel, facilities managers, network designers, estimators, inspectors, etc. need to know about fiber optics to ensure the success of their project?

The responsibility for the success or failure of any project ultimately lies with the project manager. We've seen quite a few instances of fiber optic project problems caused by improper management and many of the help calls we get at FOA indicate the manager's lack of knowledge of fiber optics. Some of the problems they call us about are amazing. An IT manager for a large metropolitan area found that the cable plant he had installed didn't work because it had 4,000 bad connectors. Another sent us OTDR traces submitted by his contractor for documentation that showed the cables were too short to test with an OTDR. In one big project, contractors subcontracted to firms that had no fiber experience who were digging up and breaking underground utilities daily.

These kinds of problems can be cured easily if the managers have some basic knowledge of fiber optics. They do not need a typical FOA fiber optic training course because those courses are based on KSAs – the knowledge skills and abilities needed by installers. What they need is just a basic understanding of fiber optic network design, installation, testing and operation.

Who Is a "Manager"?
The manager may be the supervisor of a crew of installers building the network, of course, or the manager of a contracting company. There is the communications or IT manager who works for the owner of the network, specifies the communications requirements, and has responsibility for the operation of the network after construction. The buildings or facilities manager overseeing the locations where the project is installed may be involved in its installation, operation, and maintenance. In some cases, it would include the inspector overseeing the construction and approving it. In this category, we include anyone who is involved with the network and has responsibilities that include the fiber optic network itself.

Here at the FOA, we get lot of calls from those kinds of people asking questions that show they need to know (and want to know) more - at least enough to make intelligent decisions regarding the project that affect its success. This article will cover what we think the bosses need to know based on what they have asked us.

The Basics - What Does A Manager Need To Know?
Fiber optic communications is quite simple. Instead of sending signals as pulses of electricity or radio waves, fiber optics uses pulses of light transmitted down a hair-thin ultra-pure strand of glass. Cables holding tens, hundreds or even thousands of fibers can be run underground, aerially on poles or even under water. Construction of a fiber optic cable plant is similar to that of any other cable and there are thousands of trained and FOA-certified techs available to build fiber optic networks.

Managers need to know the basics, the jargon, and how to communicate with suppliers, contractors, and installers. Forget the physics and optics - not even installers need to know the technology that makes fiber optic communications possible. Managers do need to learn about fiber optic components like the types of fibers (singlemode or multimode) used in various networks to ensure the proper ones have been chosen for the installation. We prevented a manager recently from ordering tens of miles of outside plant cable with the wrong fiber - multimode not singlemode. Hopefully a salesperson, distributor or manufacturer would have questioned his choice but if not, he would be stuck with a large amount of virtually worthless cable.

They should also learn about cables and their applications. We've seen specs for direct burial armored cables that were to be pulled through conduit and non-armored cable designed into a project for direct burial. We've seen indoor cable specified for outdoor installation and outdoor cable specified for premises installation. You must know what the proper cable choice for the installation is.

Fiber optic connector compatibility is another important issue. Twice recently I have been asked by managers about the difference between two types of connectors - PC (physical contact) and APC (angled physical contact) connectors - and whether they are compatible. They certainly are not and they may be damaged by mating to the wrong type. But try to find that advice on a manufacturer's or distributor's website - they expect everyone to know that already.

Those can be expensive mistakes! A few minutes learning the basics from books or online at Fiber U or the FOA website can answer those questions and prevent some big problems. Or just call us at the FOA – that's what many people do.

Don't believe the classic "myths of fiber optics." I once jokingly threatened physical harm to the new editor of one magazine I write for if he ever published another article that said "fiber optics is fragile because it's made of glass, is much more expensive than copper cables and is very hard to install."

Let's kill off those myths once and for all. The pure glass in optical fiber is many times stronger than steel and fiber optic cable is much more flexible than coax or twisted pair copper cable. Even 30 years ago, fiber had the bandwidth and distance advantages that made communications over fiber optics cost only s few percent as much as over copper or microwave radio. Today we can put almost one million times more communications over fiber than back then. And finally, there are more than 100,000 skilled installers who have installed millions of miles of fiber and will attest to the fact that it's just another skill to learn.

The Design
It is at the design stage that the manager has the most important role in the success of a fiber optic project. This is not a time to delegate without oversight. The manager must be able to evaluate options presented and make decisions based on the input of many others.

If someone who works for you is designing a fiber optic network, they need to know whether it provides the communications capacity you need for today and over its projected lifetime. Are there enough fibers for spares and future expansion? Can the network support drops to new user locations? Has the network been designed optimally for both performance and cost? Are all the components chosen appropriate for the network? Is the network secure and are you prepared to restore outages? One good test is to create a scope of work (SOW) and send out a request for proposal (RFP) to some experienced contractors for comments.

FOA has a complete textbook on fiber optic network design, but the basics are summarized in the FOA Guide online.

Construction And Installation

Fiber optic cable plants can be installed outside (called "OSP" for outside plant) or indoors (called "premises"). The OSP cable plant can be installed underground, aerial or under water. All have various techniques that can be chosen depending on the geography of the route or local requirements, for instance that all cables must be placed underground. Premises cabling is often a mix of fiber optics and copper cabling. It will be covered by codes like the NEC to ensure safety for those inside the building.

The FOA Guide has a section on Construction and another on Installation.

The Contractor

How do you evaluate contractors? The top of the list of requirements is experience in similar jobs backed by great references. Are their designers, managers and installers properly trained and certified? How much personnel turnover do they have? What's their plan for on-the-job training (OJT) for new recruits? Are they fully equipped for the job? What other jobs are they qualified for? Electrical construction and fiber optics are often done by the same contractor - although by different divisions of the same company - and may yield more efficient construction when electrical services are required in communications facilities.

If the contractor is chosen in a bid process, don't blindly choose the lowest bidder. Include in the RFQ (request for quotation) requirements for the bidders to include lots of information about the company that will allow evaluation of their ability to complete the job properly, including company history, personnel, structure, financial history, worker credentials, experience and of course references.

We've seen jobs go to the lowest bidder where the contractor installed thousands of splices and connectors improperly, submitted erroneous test data, got paid and disappeared, leaving the network owner holding the bag. In another case of improper installation, the contractor went bankrupt when forced to redo the job correctly.

Choosing a fiber optic contractor is another topic covered in the FOA Guide.

Finding Workers And Workforce Development

After financing, finding enough qualified workers to build your project is probably the next biggest problem. The fiber optic industry is growing so fast that there is a big shortage of qualified installers and finding good installers is very difficult. These people can also be expensive, so some contractors subcontract work to anybody, qualified or not. Several projects got caught using landscapers to install fiber after they cut several fiber optic cables already in the ground in one city and laid cable on the top of the ground in tall grass – not buying it at all – in another city. A top priority should be finding qualified workers or training them yourselves.

See "Workforce Development" below.

Evaluating The Quality Of An Installation

If the contract covers both electronic equipment and fiber optic cable plant, the number one concern is if the communications system works as planned. Under any circumstances, the quality of the fiber optic cable plant needs to be evaluated independently. Every step of the way should be documented and inspected to ensure that the network was installed in a "neat and workmanlike manner." The installation needs to be completely tested to confirm it meets the design goals and documentation of the test results presented along with the other project documentation. Fiber optic testing is a complex process that requires a trained and experienced tech to perform properly.

Documentation

Too many networks have inadequate documentation, insufficient to evaluate the installation, allow moves, adds and changes (MACs) or restoration in an emergency. Many managers and installers think the documentation is created after the network is built, but that's completely wrong. Network documentation starts when the idea of the network is conceived, evolves through the design, creation of the scope of work (SOW), RFP and RFQ (request for quote), installation and testing. Documentation should be one of the legal requirements of the contract for network installation. The installer should get the final payment only after they submit all the documentation required, not before.

Documentation must include the route of the cable plant and the type of installation (aerial, underground, etc.) and location of every component of the fiber optic cable plant including cables, splices, terminations, pedestals, manholes/handholes, etc. The documentation must include the path of every cable, every fiber in the cable (with color codes) and the test results from testing each fiber. If that sounds like a lot of work and a lot of data, it is, but that's what's necessary to determine what has been installed and if it was installed according to the plans. That data will be invaluable when changes need to be made to the cable plant or restoration must be done in event of a cable break.

There are software aids for documentation. Geographic information systems (GIS) are now widely used for for both aerial and underground utility locations and can be used to also locate the fiber optic cable plant. Other software for documenting the cable plant are available or one can create their own with database or spreadsheet programs. For premises cabling, software similar to that used for designing electrical systems are readily available and may be useful for some OSP applications. They offer the advantage of helping with estimating too.

There is more information on project paperwork from the FOA Guide. And when do you know the cable plant installation is complete? There is a page on project "deliverables."

Operating A Fiber Optic Network

Everyone who converts to fiber learns fast that fiber needs virtually no maintenance. Fiber should be installed, tested, locked up and forgotten unless you need to modify the network or repair damage. Most damage to the network is caused by poorly trained techs working with cables they don't understand. Another major problem is damage outside your control - underground cables suffering what we in the industry call "backhoe fade," or for aerial cables what a utility out West referred to as "target practice".

Emergency Restoration

Like any other problem, restoring a fiber optic network failure is easier if you plan ahead. If you have damage, the most valuable tool you have for restoration is all the documentation on the network. With that you know exactly where the cable plant is installed and troubleshooting test results can be compared to the fibers when installed. Leftover components like spools of cable, splice closures or other hardware should be kept, stored with the documentation for use in restoration. And, of course, you need trained crews on 24/7 call, who have the skills to track down problem and fix them. If you don't have your own personnel who can do this, have a contract with someone who can and will respond quickly.

Getting Up To Speed

How does a manager learn all this? You can learn by experience, of course, although that's often a painful way to learn. If your personnel are being trained, take a course with them. If you want to learn on your own, there is plenty of information on the FOA website and free self-study programs at Fiber U that can help you understand fiber optic project management.

If You Are Considering DIY FTTH Project, Here Are Things To Remember

Legality: If you are in one of 18 US States, your state legislators have passed laws written by lobbyists for incumbent service providers that prevent municipalities, other governments, or coops from becoming ISPs. By the time you read this, all this may have been negated by local or Federal laws.

Expect A Fight: Most independent FTTH projects, especially those proposed by municipalities or coops, will be opposed by the incumbent service providers. Experience has shown that they have no problems spending lots of money opposing the project and fighting dirty. Stay calm, tell your story, people are learning about what to believe.

Uniqueness: Like most fiber optic networks, every FTTH installation is unique. It must be designed for the location it is to serve and choices on components and installation methods should be optimized for the system. Construction and installation methods may include every type of OSP installation. Suppliers familiar with FTTx can advise customers on what others have done to make installations simpler, easier and less expensive. Most systems prefer to use as many factory-made components as possible as they are generally less expensive than doing the same work in the field. New installation methods should be considered as well to reduce costs.

Consultants: Be wary of consultants. Consultants can be extremely valuable in designing a FTTH system, as long as they have relevant experience, are up to date on new components and techniques and are highly recommended by previous clients. Unfortunately, we have seen problems with consultants, including over-designed networks with costs much higher than necessary, installation practices recommended that were unnecessary or ignore newer technology, systems designed around components that were higher performance (and price) than necessary, and in one case a consultant took the clients payment, went away for a year and came back with an admission that they could not design the network (but they kept the consulting fees.)

Contractors: As with any fiber optic project, the quality of the installation depends on the quality of the installer. Look for contractors with knowledge, experience and references. And preferably relevant certifications like the FOA CFOT. Be especially wary of subcontractors. Any subcontractors should have equal qualifications and be approved by the network owner. We have seen landscape contractors with no fiber training used as subcontractors for cable plant installation - one cut several cables to buildings that had been installed by a member of the FOA advisory board!

Call Before You Dig! Every day some major fiber optic cable is cut by a contractor. The jurisdiction issuing permits should help you with locating other buried utilities. There is a service that helps you locate underground utilities that may be in your construction path. See the FOA web page on Digging Safely.

What Fiber Do You Already Have? Before you design or install a new fiber optic cable plant, inventory the fiber you have already and/or negotiate to lease fiber where others have cables with dark (unused) fibers. Also talk to other organizations who may need communications to see if they want to share costs or lease dark fibers or communications links from you. Cities, counties and states need fiber. Utilities need fiber. Fire and life safety organizations need fiber. Traffic departments need fiber. Cellular companies really need a lot of fiber.

What Other Services Can Share The Fiber?   Consider what other services than FTTH you can carry on your fiber optic cable plant - cellular backhaul, traffic systems, security/surveillance systems, leased fiber, etc. to generate additional revenue. A few years ago a large American city sent out a RFP (request for proposal) for an urban FTTH network. The document dealt strictly with FTTH to connect the city's citizens with fiber and ignored all the other services the city had that already used or needed fiber - city communications, security/video surveillance, intelligent traffic management, public transportation communications, wireless networks(small cells and 5G), utility communications, etc., etc., etc.

Dig Smart - Dig Once: This same document also covered the difficulty of urban installation - digging up streets already filled with underground utilities, limited space for pedestals, few options for aerial cable and other issues that are typical problems for urban fiber installation. No mention of "Dig Once" to make future installations easier. Share fibers. Use spare fibers. Use additional wavelengths in current fiber. Consider all the alternatives. Plan ahead - future proof is a myth, but one can make certain decisions that will make the future easier.   If you are considering using FTTH design software, ask to talk to customers who have used it. Determine what you need to know first in order to use it, e.g. GIS data on every utility pole, manhole or handhole, subscriber location, etc. and how much training it takes to become proficient. Will you use your personnel or hire outsiders, and how do you evaluate them.

Cost Savings: Fiber optic cable and components are not expensive, but labor is. Saving money on components may look good in first analysis, but more savings will come from optimized designs and efficient installation practices. More experienced contractors are more efficient and may save costs by their speed and efficiency. And design for the future - if you dig a trench for anything, not just fiber but any underground utility, bury a number of fiber ducts for future use, install cables with more fibers than you need - lots more - fiber is cheap, installation is expensive. The program is called "Dig Once."

Take Rates Are Important: "Take rates" for new FTTH networks – the percentage of homes passed that become subscribers - vary from low to high, depending on the satisfaction with the current ISP (Internet service provider.) When Google Fiber started in Kansas City, the take rate was high because the current service was bad, but in later cities when the local ISPs knew they were coming and improved their service and/or lowered their prices, the take rate was lower. Competition tends to drive take rates and take rates determine the economics of the system, Know your competition. Offering gigabit services are often the top selling point of FTTH. Every GPON network is a gigabit network, but subscribers can opt for slower speeds at lower costs.

Workers And Workforce Development

Fiber optics, like any fast-growing technology, needs well-trained workers and an FTTH project is no exception. The FOA has partnered with a number of organizations to develop guidelines for hiring and retaining competent workers. Here's what we have learned:

First, here are some important questions to ask when hiring installers and other personnel:

Who hires your installers? This seems like an obvious question but when you move down from the Contractor who hires the subcontractor, it is the subcontractor (or the sub-sub subcontractor if you are planning a large project) who hires the installers and other workers. To complicate things even further, sometimes it is a placement agency that may provide temporary workers for your job.

How do you ensure that these workers are properly trained? We have seen the results of poorly trained installers all too often. Examples are given in other sections of this book. The only way to combat this is to identify and communicate directly with the actual hiring manager - contractor? subcontractor? placement agency? You might be surprised if you do some research how this is delegated to someone down the "chain of command".

Here are some questions to ask:
What qualifications are required? Is there a job description? Does the job description accurately reflect the technical requirements of the job. Read the contract carefully – is experience and/orcertification required for the techs?

Along with technical skills required for the job, are they checking for "life skills" such as basic math skills and a 10th grade reading level as a minimum and basic computer knowledge? Remember how extensive the use of computers are in the field now.

Is the candidate evaluation done completely online or is there a face-to-face meeting before the hire?

How are the candidates' attitude and behavior assessed? Are they checking references to determine items such as adaptability, reliability, integrity, having good judgment? Judgement is important especially when safety is critical.

Are you reaching out to local technical schools? Contact local community colleges and tech schools about your worker needs and the job opportunities available to their graduates. If your local schools do not have fiber optic or telecom programs, what technical subjects do they teach that may produce good candidates? The FOA has Approved Schools all over the country and many have advisory boards seeking advice from local businesses on the need for specialized jobs and training.

If local colleges or technical schools are interested in teaching a fiber program for telecom or IT, FOA has everything they need to start a program quickly. Many of these schools have Workforce Development departments which have access to US Dept of Labor and state job training dollars that can fund a program.

You can develop your own workforce – train and promote from within using structured OJT – online learning combined with learning on the job. As the shortage of skilled workers become more acute, contractors are going to have to train their own people. The reality of the workplace is that most training is "on the job" (OJT).

OJT should be more than giving someone a set of tools and have them follow someone around the worksite to learn. The FOA has a program we call OJT to Cert where you can use Fiber U free online training with a designated supervisor who sets a timetable for achieving specific learning goals and conducts ongoing evaluations for an "OJT" candidate. FOA certification could be achieved in one year.

Hiring new people is not the only issue facing contractors. Employee retention is a growing problem - contractors should not overlook the needs of existing employees to stay current with new technologies and develop necessary skills at all stages of their working career. Most of the workforce is now familiar with online learning like the FOA offers on Fiber U that can lead to certification. For instance, the FOA has a Direct Work to Cert certification program for experienced techs.

What Makes A Successful Fiber Optic Project?
People call FOA for advice all the time. Most of the calls deal with technical questions about products, installation and testing. But in one call; a manager who was starting to plan a fiber optic project wanted advice on how to proceed. It was a long call! His basic question was "What does it take to have a successful fiber optic project?" We responded with 4 words: financing, commitment, expertise and patience. (This section is repeated from the introductory section on FTTH because it's important for the designer and managers of a FTTH project.)

Financing: The story goes that someone asked Neil Armstrong what he was thinking about while sitting on top of the rocket ready to launch Apollo 11 to the moon. "Every part was made by the lowest bidder," was supposedly his reply. (The same quote has been attributed to most early astronauts!)

Fiber optics are not necessarily expensive; in fact, fiber has been used so widely because it is the least expensive communications medium in virtually all projects. But fiber optic projects may require a lot of construction which makes the project expensive. Like all other projects, it never pays to cut corners. Planning and running the project properly is what saves money, trying to cheapen the project. Not all jobs should go to the lowest bidder, unless they meet all the criteria for a qualified bidder. Likewise, one needs to ensure that when a project starts, there are funds available to complete the job properly, including some extra for unplanned changes or modifications.

Commitment: Just like having sufficient finances to compete the project, one needs a commitment to finish the job once it is started. Changes of management or changes in governments often lead to confusion or even modifying a project in midstream. There is nothing wrong with making changes based on what learns as the project progresses; it may even involve greater efficiency or cost savings, but arbitrary changes may jeopardize the project's timetable, completion or even its usefulness.

If the project is under the auspices of a government entity, changes in administration or management that causes changes in a project will invariably make it more expensive and may jeopardize the success of the entire project. Ideally, the personnel who propose, design and plan the network should see it to completion.

Expertise: Fiber requires expertise and experience. It's obvious the installers need to know what they are doing, but in reality, so must the managers who work for the organization that is contracting for the work. There are many instances of projects where the managers signed off on the project when it was incomplete or improperly installed. The only way to properly manage a project is to understand every aspect of it well enough to know if it is being done properly and when it is actually complete.

Planners, designers, contractors and installers should all be trained and certified as well as being experienced with good references. That holds doubly so for consultants. In many places, to be a consultant or cabling contractor means little other than registering as a business and advertising your services. Some of the problems we've seen with outside services, include consultants who took contracts, spent time on a project, then told the customer they could not help them with the project, but kept the money.

We have seen contractors doing shoddy installations, ruining expensive fiber optic cable during pulling and leaving jobs half done but getting paid because the customer knew no better. One of the biggest problems is subcontractors. A contractor with good credentials gets the job but subcontracts some of the work to a contractor who will do the work at a lower price, but does not have the training or experience (or motivation) to do it right. In your contract with an installer, we recommend a clause giving the project manager responsibility for evaluating and approving all subcontractors.

The manager must know better to prevent problems. FOA also has pages on what the manager needs to know. See the FOA Guide.

Patience: From concept to acceptance, a typical OSP fiber project can take 2-5 years and a premises project 1-2 years. It depends on the size of the project, the time to properly design it, create project paperwork, get permits, buy components, hire contractors and properly install it. Proper workmanship takes time and is not easily rushed. Saving time generally means cutting corners and that is often the cause of the problems encountered. Take your time, plan, design, select, install, test and document your network properly.

And by the way, "future proofing" is a myth! Who would have known in 1990 how ubiquitous the Internet would be today? How reliant we could be on smartphones other mobile devices? How many workers would be working remotely or using videoconferencing for meetings? Technology moves too fast and is too disruptive for anyone to make reliable predictions. The IBMer who developed MRP - the original company organizational software - used to tell everyone, "A forecast is wrong from the moment it is made." Plan for the future, but assume you will upgrade, change directions, etc. driven by new tech and changes in the world around us.

References:

Additional References For Training and Study For Additional Knowledge or FOA Certification

As with any fast-moving technology, keeping abreast of the latest technology, techniques and products can be a daunting task. Here are some references which will assist you.

FOA Websites
The FOA website, www.foa.org, has a special section of the Online Fiber Optic Reference Guide (www.foa.org/guide) with ~1,000 pages on fiber optic technology and practice.

Fiber U
The FOA has also created an online learning site, Fiber U at www.fiberu.org, that offers free online courses for self-study.

FOA Textbooks
The FOA has published other Reference Guides that are the references for FOA certifications.

The FOA Reference Guide to Fiber Optics is a general reference guide for fiber optics and the basic study guide for the CFOT certification.

The FOA Reference Guide to Premises Cabling is a reference guide for copper and fiber optic cabling and wireless as used in indoor applications and the basic study guide for the CPCT certification.

The FOA Reference Guide to Outside Plant Fiber Optics is a reference guide for fiber optic cabling as used in outdoor applications and the basic study guide for the CFospT certification.

The FOA Reference Guide to Fiber Optic Network Design is a guide to the design of fiber optic projects, from concept to operation.

The FOA Reference Guide to Fiber Optic Testing is a comprehensive guide to testing fiber optic components and networks.

All FOA textbooks are available for purchase from Amazon and most booksellers.

FOA/NECA 301 Fiber Optic Installation Standard is a ANSI standard covering the installation of fiber optic cable networks written by the FOA in conjunction with the NECA NEIS standards series. It is available for free downloads from the FOA website.

The Fiber Optic Association

Fiber To The Home (FTTH) Handbook

The Fiber Optic Association, Inc.
Telephone: 1-760-451-3655 Fax 1-781-207-2421
info@foa.org   www.foa.org

www.ingramcontent.com/pod-product-compliance
Lightning Source LLC
Chambersburg PA
CBHW081435220526
45466CB00008B/2400